LIFE CYCLE COSTING

LIFE CYCLE COSTING

Techniques, Models and Applications

by

B.S. Dhillon
Department of Mechanical Engineering
University of Ottawa
Ontario, Canada

GORDON AND BREACH SCIENCE PUBLISHERS
New York London Paris Montreux Tokyo Melbourne

Gordon and Breach Science Publishers

Post Office Box 786
Cooper Station
New York, New York 10276
United States of America

Post Office Box 161
1820 Montreux 2
Switzerland

Post Office Box 197
London WC2E 9PX
England

Private Bag 8
Camberwell,
Victoria 3124
Australia

58, rue Lomond
75005 Paris
France

3-14-9, Okubo
Shinjuku-ku, Tokyo
Japan

Library of Congress Cataloging-in-Publication Data

Dhillon, B.S.
Life cycle costing.

Includes bibliographies and index.
1. Engineering economy. 2. Cost control.
3. Product life cycle. I. Title.
TA177.7. D35 1988 658.1′552 88-21238
ISBN 2-88124-302-9

CONTENTS

PREFACE

Due to various reasons, the equipment procurement process, in general, is experiencing a new trend. This trend is based on the principle that a newly procured product must be supported for its total life years. In several cases, the product acquisition cost is less than the support cost over the life cycle years. Depending on the equipment type, the ownership cost over the life span may vary from 10 to 100 times the acquisition cost. Therefore, in the purchase of new major equipment, procurement management, particularly in the military, examines the total life cycle cost of the product rather than just its initial cost.

Over the years, many advances in life cycle costing have resulted in a large amount of published literature—witness the appendix of this book. This book covers these developments and recent related progress in the field. Topics of current interest are treated in such a manner that the reader needs no previous knowledge to understand the contents. Four chapters of the book provide background information on life cycle costing and on related areas. The source of most of the material presented is given in references for those readers who want to delve deeper into a specific area. The text cites over 1150 references, most listed in the Appendix, as well as some at the end of each chapter. In addition, the book contains several examples and their corresponding solutions. Throughout the text, emphasis is made on the structure of concepts rather than on minute details.

The book contains twelve chapters and the Appendix. Chapter 1 briefly discusses life cycle costing history, definitions, and scope of the book. It lists related periodicals, conference proceedings, and books. Chapter 2 reviews basic economic concepts useful for life cycle costing. Life cycle costing basics are described in Chapter 3.

Chapter 4 is devoted to life cycle models. These models are separated into two major categories: general life cycle cost models and specific life cycle cost models. Ten general and thirteen specific life cycle cost models are presented. Chapter 5 presents several cost estimation techniques. Fifteen techniques are described and sources for cost data are listed.

Chapter 6 is divided into three parts: manufacturing, quality, and maintenance costs. All three topics are covered in considerable depth. Chapter 7 is devoted to warranties. This chapter covers topics such as general warranties, reliability improvement warranties, and warranty-related cost models. Cost estimation models are discussed in Chapter 8. This chapter contains a total of twenty-five models ranging from estimating the equipment development cost to cost of centrifugal pumps.

Computer hardware and software costing is discussed in Chapter 9. Basically, the chapter is divided into two major parts: computer hardware costing and computer software costing. A list of software cost databases is given. The next two chapters, Chapters 10 and 11, deal with life cycle costing in the aircraft industry and life cycle costing of energy systems and on-surface vehicles, respectively. Chapter 10 covers topics such as cost drivers, aircraft life cycle cost, and models employed in estimating life cycle cost for United States Air Force aircraft systems, and Chapter 11 presents several life cycle cost models concerned with energy systems ranging from circuit breakers to water heaters. In addition, life cycle cost models for urban rail, automobile, cargo ship, etc., are presented.

Chapter 12 is completely devoted to reliability. Several important areas of reliability directly or indirectly useful for life cycle costing are covered. Some of those are reliability networks, maintainability measures, and availability.

This book should be useful to procurement, maintenance, reliability, design, maintainability, systems, chemical, electrical, computer, mechanical, civil, electronics, quality control, and aeronautical engineers, as well as senior students of engineering and business administration. Administrators, cost analysts, researchers, academics, and others should find it a valuable reference source.

ACKNOWLEDGMENTS

I wish to thank MCB University Press for granting permission to duplicate some of the references listed in the Appendix from an article by Y. Gupta and W.S. Chow, "Twenty-Five Years Life Cycle Costing—Theory and Applications: A Survey", *The International Journal of Quality & Reliability Management*, Vol. 2, 1985, pp. 51–76. I am indebted to my relatives and friends for their interest and encouragement during moments of need. I thank my wife, Rosy, for typing the original manuscript and for her ever present assistance during the preparation of the manuscript.

CHAPTER 1

Introduction

1.1 Brief Life Cycle Costing History

It was almost a quarter century ago when wide dissemination of the term life cycle costing was given in a report [1] entitled "Life Cycle Costing in Equipment Procurement". This document was the result of a study conducted by the Logistics Management Institute, Washington, D.C. for the Assistant Secretary of Defense for Installations and Logistics. As a result of this report, a series of three guidelines [2–4] for life cycle costing procurement was published by the United States Department of Defense. These guidelines were entitled "Life Cycle Costing Procurement Guide (interim)", "Life Cycle Costing in Equipment Procurement-Casebook", and "Life Cycle Costing Guide for System Acquisitions (interim)". In 1971, the Directive 5000.1 [5] entitled "Acquisition of Major Defense Systems" was issued by the Department of Defense. This directive established the requirement for life cycle costing procurement for major defense systems acquisitions.

A project entitled "Life cycle budgeting and costing as an aid in decision making" was initiated by the United States Department of Health, Education, and Welfare in 1975 [6]. In 1974, the state of Florida formally adopted life cycle costing and in 1978, the United States Congress established the National Energy Conservation Policy Act. This Act requires every new federal building be life cycle cost effective.

Since 1974, many states in the United States have passed legislations making life cycle cost analysis mandatory in the planning, design, and construction of state buildings. Some of these states are New Mexico, Alaska, Maryland, North Carolina, and Texas. Over the years many other events have taken place in the history of life cycle costing and a large amount of published literature on the subject have appeared. References [7, 8] list a large volume of published literature on life cycle costing. An extensive list of publications on life cycle costing and related areas is presented in the Appendix.

1.2 Periodicals and Conference Proceedings

Information on life cycle costing can be found in various journals, conference proceedings, books, and reports. To the best of author's knowledge there are no periodicals or annual conferences specifically on the subject of life cycle costing. However, there are several journals and conference proceedings which time to time publish articles on the subject in question. Some of those journals and conference proceedings are listed below:

Journals

i) Microelectronics and Reliability: An International Journal
ii) IEEE Transactions on Reliability
iii) Defense Management Journal

Conference Proceedings

i) Proceedings of the Annual Reliability and Maintainability Symposium, USA.
ii) Proceedings of the Annual International Reliability, Availability, and Maintainability (Inter-RAM) Conference for the Electric Power Industry
iii) Proceedings of the Annual American Society for Quality Control (ASQC) Conference

1.3 Books

So far there has been only a limited number of books written on life cycle costing. Those books are listed below:

i) M.R. Seldon, Life Cycle Costing: A Better Method of Government Procurement, Westview Press, Boulder, Colorado, 1979.
ii) R.J. Brown, R.R. Yanuck, Life Cycle Costing: A Practical Guide for Energy Managers, The Fairmont Press, Inc., Atlanta, Georgia, 1980.
iii) B.S. Blanchard, Design and Manage to Life Cycle Cost, M/.A Press, Portland, Oregon, 1978.
iv) M.E. Earles, Factors, Formulas, and Structures for Life Cycle Costing, Second Edition, Eddins-Earles, Privately Published, Concord, Massachusetts, 1981.

v) A.J. Dell'isola, S.J. Kirk, Life Cycle Costing for Design Professionals, McGraw-Hill Book Company, New York, 1981.

A chapter on life cycle costing is given in the following books:

i) B.S. Dhillon, Reliability Engineering in Systems Design and Operation, Van Nostrand Reinhold Company, New York, 1983, pp. 210–238.
ii) B.S. Dhillon, H. Reiche, Reliability and Maintainability Management, Van Nostrand Reinhold Company, New York, 1985, pp. 214–235.
iii) H. Reiche, Life Cycle Cost, in Reliability and Maintainability of Electronic Systems, Edited by J.E. Arsenault and J.A. Roberts, Computer Science Press, Potomac, Maryland, 1980, pp. 3–23.

1.4 Definitions

This section presents selected definitions directly or indirectly related to life cycle costing [9–14].

Life Cycle Cost: the sum of all costs incurred during the life time of an item, i.e., the total of procurement and ownership costs.

Procurement Cost: the total of investment or acquisition costs (recurring and non-recurring).

Ownership Cost: the sum of all costs other than the procurement cost during the life time of an item.

Recurring Cost: the cost which recurs periodically during the life of a project.

Reliability: the probability that an item will carry out its mission satisfactorily for the desired period when used according to specified conditions.

Failure: the termination of the ability of an item to carry out its stated mission or function.

Maintenance Cost: the labor and materials expense required to maintain item(s) in suitable use condition.

Manufacturing Cost: the sum of fixed and variable costs chargeable to the production of a specified item.

Maintainability: the probability that a failed item will be restored to its satisfactory operational state within a specified total downtime when maintenance action is started according to stated conditions.

Redundancy: the existence of more than one means for carrying out a stated function.

Cost Model: an approach, based on technical and programmatic parameters, for computing concerned costs.

Nonrecurring Cost: the cost that is not repeated.

Maintenance: all scheduled and unscheduled actions appropriate for keeping an item in a serviceable condition or restoring it to serviceability. It includes inspection, repair, modification, servicing, remove and replace, etc.

Annuity: a series of equal payments, at equal intervals.

Cost: the amount of money paid or payable for the acquirement of services, materials, or property.

Downtime: the total time during which the product (or item) is not in a condition to carry out its stated function or mission.

Failure Rate: the number of failures of a product per unit measure of life (e.g. hours).

Active Redundancy: the term used when all redundant items are operating simultaneously.

Battery-Limits: a geographical designation for the process area of the project proposed.

Cost Estimating Relationship: an equation relating cost as the dependent variable to one or more independent variables.

Mean Time to Repair: the mean time needed to repair a product (or an item).

Repair Cost: the cost of restoring an item or a facility to its original condition or performance.

Compound Amount: the future value of money invested or loaned at compound interest.

1.5 Scope of the Book

Nowadays, life cycle costing is receiving increasing attention in government setups, industry, etc. Most of the information on life cycle costing is available in the form of technical papers or reports and there are only a limited number of books specifically dealing with life cycle costing. All of those books appeared during the period from 1978–1981. Since 1981, to the best of author's knowledge not a single text appeared on the topic but there have been many articles and reports. Professionals and others involved in life cycle costing need up to date information on the subject in question and generally face a great deal of difficulty. This book is an attempt to fulfill the current need. The book is written after reviewing the available literature on life cycle costing.

Therefore, every attempt was made to cover important past and current issues in life cycle costing. Previous knowledge is not generally necessary to comprehend the contents, since two chapters on basic economics and reliability theory are provided to give sufficient background. This book will find use in many disciplines and will be useful to reliability, maintainability, and quality control engineers, managers, procurement specialists, cost analysts, project, maintenance, design, system, computer hardware and software, chemical, electrical, electronic, civil, and mechanical engineers, researchers and university level teachers, and senior students of engineering and business administration.

1.6 Summary

This chapter presented a brief history of life cycle costing. A list of journals and conference proceedings publishing articles on life cycle costing is given along with books concerned with the topic in question. Several definitions directly or indirectly related to life cycle costing are presented. The scope of the text is discussed.

1.7 Problems

1. Write an essay on the history of life cycle costing.
2. Define the terms given below:

   i) Life cycle costing
   ii) Interest rate
   iii) Disposal cost
   iv) Economy
   v) Development cost
   vi) Mean time between failures
   vii) Cost drivers
   viii) Warranty
   ix) Design to cost
   x) Direct cost

1.8 References

1. "Life Cycle Costing in Equipment Procurement", Report No. LMI Task 4C-5, Prepared by Logistics Management Institute (LMI), Washington, D.C., April, 1965.

2. "Life Cycle Costing Procurement Guide (interim)", Department of Defense Guide No. LCC-1, United States Department of Defense, Washington, D.C., July 1970.
3. "Life Cycle Costing in Equipment Procurement-Casebook", Department of Defense Guide No. LCC-2, United States Department of Defense, Washington, D.C., July 1970.
4. "Life Cycle Costing Guide for System Acquisitions (interim)", Department of Defense Guide No. LCC-3, United States Department of Defense, Washington, D.C., January 1973.
5. "Acquisition of Major Defense Systems", Department of Defense Directive No. 5000.1, United States Department of Defense, Washington, D.C., July, 1971.
6. M.E. Earles, Factors, Formulas, and Structures for Life Cycle Costing, Published by Eddins-Earles, 89 Lee Drive, Concord, Mass. 01742.
7. B.S. Dhillon, Life Cycle Cost: A Survey, Microelectronics and Reliability, Vol. 21, 1981, pp. 495–511.
8. B.S. Dhillon, Reliability Engineering in Systems Design and Operation, Van Nostrand Reinhold Company, New York, 1983, pp. 210–238.
9. K.K. Humphreys, P. Wellman, Basic Cost Engineering, Marcel Dekker, Inc., New York, 1987.
10. R.D. Stewart, R.M. Wyskida, Cost Estimator's Reference Manual, John Wiley & Sons, New York, 1987.
11. M.R. Seldon, Life Cycle Costing: A Better Method of Government Procurement, Westview Press, Boulder, Colorado, 1979.
12. R.J. Brown, R.R. Yanuck, Life Cycle Costing, The Fairmont Press, Inc., Atlanta, Georgia, 1980.
13. K.K. Humphreys, Project and Cost Engineers' Handbook, Marcel Dekker, Inc., New York, 1984.
14. "Aircraft Engine Life Cycle Cost", Document No. SP-721, Society of Automotive Engineers, Inc., Warrendale, Pennsylvania, 1987.

CHAPTER 2

Life Cycle Costing Economics

2.1 Introduction

The discipline of economics plays an important role in life cycle costing. In order to calculate life cycle cost of items various types of economics related information is required.

Life cycle costing requires that future costs have to be calculated by taking into consideration the time value of money. This is due to the fact that same sum of money received or spent at various different points in time will have different values. More specifically, the sum of money in today's dollars, usually, will not have the same value a year later and thereon. Taking this one step further, one may state that the future value of present dollars will be greater because of earned interest or smaller because of inflation. Similarly, the present value of a future sum of money would be generally less. In life cycle costing, the future costs such as operation and maintenance have to be converted to their appropriate values before adding them to the item's procurement cost. There are a number of formulas developed in economics to convert money from one point of time to another. In life cycle costing studies such formulas are indispensible.

This chapter presents several aspects of economics useful for life cycle costing.

2.2 Early History of Interest

The concept of interest is not new as many of us may believe. Just like taxes its history goes back thousands of years. The early history of man reveals that it existed, in Babylon, two thousand years before the birth of Jesus Christ (i.e. 2000 B.C.). In those days interest was paid on borrowed commodities (e.g. grain) in form of grain or other means. By the year 575 B.C., the concept of interest was so well developed that led it to the formation of a company of international bankers with home offices in Babylon [1]. The firm derived its income by charging high interest rates on its money used for financing international trade.

2.3 Simple and Compound Interest

The simple interest is the simplest form of interest. What it means is that the interest is paid only on the original sum of money borrowed, and not on the accrued interest. In the case of simple interest, the total interest earned or paid is given by

$$TI = (PA)(j)(m) \tag{2.1}$$

where

TI is the total interest
m is the interest periods (e.g. years)
j is the interest rate (per specified period)
PA is the principal amount (i.e. borrowed or lent)

The total amount, TA, of money at the end of m years can be obtained from the following equation:

$$TA = PA + TI \tag{2.2}$$

Substituting Equation (2.1) into Equation (2.2) yields

$$\begin{aligned} TA &= PA + (PA)(j)(m) \\ &= PA(1 + jm) \end{aligned} \tag{2.3}$$

Example 2.1

A person loaned \$2,000 for the period of 4 years at an annual simple interest rate of 15%. Calculate the total amount of money at the end of 4 years.

Using Equation (2.1) and the given data, the total interest earned is

$$\begin{aligned} TI &= (2{,}000)(0.15)(4) \\ &= \$1{,}200 \end{aligned}$$

Adding the above result to the original amount \$2,000, the Equation (2.3) yields

$$\begin{aligned} TA &= 2{,}000 + 1{,}200 \\ &= \$3{,}200 \end{aligned}$$

Thus, the total amount of money at the end of 4 years will be \$3,200.

In the case of compound interest, the interest earned during each interest period is added (at the end of the period) to the principal amount and thereafter it starts earning interest itself for the remaining term of the loan or investment. In order to develop a formula for the total amount, we assume that present or principal amount, PA, is invested or loaned at compound interest rate i per interest period (usually year). In this case the total amount, TA_1, of money at the end of first interest period (e.g. year) will be

$$TA_1 = PA + (PA)i$$

$$= PA(1 + i) \quad (2.4)$$

The total amount, TA_2, of money at the end of second interest period (e.g. year) will be

$$TA_2 = (\text{original principal}) + \begin{bmatrix} \text{interest for period} - 1 \\ \text{(e.g. first year)} \end{bmatrix}$$

$$+ \begin{bmatrix} \text{original principal} \\ \text{plus interest} \\ \text{for period} - 1 \end{bmatrix} (\text{interest rate})$$

$$= PA + PA(i) + (TA_1)i \quad (2.5)$$

Substituting Equation (2.4) into Equation (2.5) leads to

$$TA_2 = PA + (PA)(i) + PA(1 + i)i$$

$$= PA + (PA)(i) + (PA)(i) + (PA)(i)^2$$

$$= (PA)[1 + 2i + i^2]$$

$$= (PA)(1 + i)^2 \quad (2.6)$$

Similarly, the total amount, TA_3, of money at the end of third interest period (e.g. year) will be

$$TA_3 = TA_2 + (TA_2)i \quad (2.7)$$

Substituting Equations (2.6) into Equation (2.7) results in

$$TA_3 = (PA)(1 + i)^2 + (PA)(1 + i)^2 i$$

$$= PA(1 + i)^2[1 + i]$$

$$= (PA)(1 + i)^3 \quad (2.8)$$

Thus, the total amount, TA_m, of money at the end of mth interest period (e.g. year) will be

$$TA_m = (PA)(1 + i)^m \tag{2.9}$$

where m is the number of interest periods. From the above equation the total amount, TA_{m-1}, of money at the beginning of mth interest period (e.g. year) is

$$TA_{m-1} = (PA)(1 + i)^{m-1} \tag{2.10}$$

The interest for the mth period (e.g. year) is

$$I_m = (1 + i)^{m-1}(PA)(i) \tag{2.11}$$

where I_m is the interest for the mth period (e.g. year). In cases when the interest is compounded more than once in a year, the given yearly interest rate, say i, must be divided by the number of times the compounding is taking place during the year. To take this factor into account, Equation (2.9) is modified to the following form [2]:

$$TA_m = (PA)\left(1 + \frac{i}{k}\right)^{mk} \tag{2.12}$$

where

k is the number of times the interest is compounded per year
i is the annual interest rate
m is the number of years
PA is the principal amount
TA_m is the total amount of money at the end of year m

Example 2.2

Assume that six thousand dollars are deposited in a bank. The annual interest rate is 12%. If the interest is compounded quarterly during a year, calculate the total amount of money at the end of the 4th year.

In this example we have $PA = \$6{,}000$, $i = 12\%$, $k = 4$, and $m = 4$ years. Thus, substituting these data into Equation (2.12) yields

$$TA_4 = (6{,}000)\left[1 + \frac{0.12}{4}\right]^{(4)(4)}$$

$$= \$9{,}628.24$$

Thus, the total amount of money after 4 years will be \$9,628.24.

Example 2.3

A person deposited $4,000 in a bank at annual interest rate of 9%, compounded annually. Calculate the total amount of money at the end of the 10th year, the interest earned during the 10th year, and the total amount at the beginning of the 10th year.

In this example, the following data are specified:

$m = 10$ years
$i = 9\%$ per year
$PA = \$4{,}000$

Using these data in Equation (2.9) yields

$$TA_{10} = (4{,}000)(1 + 0.09)^{10}$$
$$= \$9{,}469.455$$

Thus, the total amount of money at the end of the 10th year will be $9,469.455.

From Equation (2.11), the interest earned during the 10th year is

$$I_{10} = (1 + 0.09)^{10-1}(4{,}000)(0.09)$$
$$= \$781.882$$

Finally, the total amount at the beginning of the 10th year from Equation (2.10) is

$$TA_9 = (4{,}000)(1 + 0.09)^{10-1}$$
$$= \$8{,}687.573$$

2.4 Nominal and Effective Interest Rates

The nominal interest rate may simply be described as an annual interest rate without considering the effect of any compounding during the year.

On the other hand, the effective yearly interest rate may be described as the true annual interest rate, considering the effect of all compounding during the year. The effective annual interest rate can be estimated from the following equation:

$$(1 + i_{ef}) = \left(1 + \frac{i}{k}\right)^k \tag{2.13}$$

The above equation is developed by reasoning that the effective interest rate compounded once in a year generates the same amount of interest as a nominal interest rate compounded k times in a year.

Rearranging Equation (2.13) yields

$$i_{ef} = \left(1 + \frac{i}{k}\right)^k - 1 \tag{2.14}$$

The symbols used in Equations (2.13) and (2.14) are defined below:

i_{ef} is the effective annual interest rate
i is the nominal interest rate (per year)
k is the number of interest periods in a year

From Equation (2.14), it is to be noted that at $k = 1$, $i_{ef} = i$. It means the nominal (yearly) interest rate is same as the effective (annual) interest rate.

Nowadays, a number of financial institutions offer several interest periods per year. One example is daily compounding. The interest compounds continuously as k approaches infinity. In limit, if interest were compounded continuously, from Equation (2.14) we get

$$i_{ef} = \lim_{k \to \infty} \left(1 + \frac{i}{k}\right)^k - 1 \tag{2.15}$$

The right hand term is rewritten in the following form:

$$\left(1 + \frac{i}{k}\right)^k - 1 = \left[\left(1 + \frac{i}{k}\right)^{k/i}\right]^i - 1 \tag{2.16}$$

Substituting Equation (2.16) into Equation (2.15) yields

$$i_{ef} = \lim_{k \to \infty} \left[\left(1 + \frac{i}{k}\right)^{k/i}\right]^i - 1 \tag{2.17}$$

Since $\lim_{k \to \infty} (1 + i/k)^{k/i} = 2.71828 = e$, the above equation leads to

$$i_{ef} = e^i - 1 \tag{2.18}$$

Taking natural logarithms of Equation (2.18) and rearranging yields

$$i = \ln(i_{ef} + 1) \tag{2.19}$$

Example 2.4

Assume that a person deposited 15,000 dollars in a bank at a nominal interest rate of 12% compounded daily, for one year. Calculate the effective annual interest rate.

In this example, we have $i = 12\%$, $k = 365$ days, and $PA = \$15{,}000$. Using these data into Equation (2.14) results in

$$i_{ef} = \left(1 + \frac{0.12}{365}\right)^{365} - 1 = 0.127475 \text{ or } 12.7475\%$$

Alternatively, from Equation (2.18) we get

$$i_{ef} = e^{(0.12)} - 1 = 0.127497 \text{ or } 12.7497\%$$

In this case Equations (2.14) and (2.18) yielded very similar results. The effective annual interest rate may be taken as 12.75%.

2.5 Formulas for Life Cycle Cost Analysis

This section presents several formulas considered useful for performing life cycle cost analysis.

2.5.1 Formula for Finding Future Worth (Single Payment)

This formula was developed earlier in the chapter and the future worth (compound-amount) from Equation (2.9) is

$$FW = TA_m = (PA)(1 + i)^m \tag{2.20}$$

where

FW is the future worth (i.e., principal plus interest due)
PA is the principal
i is the compound interest rate per period
m is the number of interest periods

2.5.2 Formula for Finding Present Worth (Single Payment)

From Equation (2.20), the present worth of a future sum of money is

$$PW = PA = \frac{FW}{(1 + i)^m} \tag{2.21}$$

where PW is the present worth.

Example 2.5

Assume that the total maintenance cost of an equipment at the end of its seven year operation will be \$80,000. More specifically, this is the amount which will occur seven years from now. The estimated annual compound interest rate is 10%. Compute the present worth of the 80,000 maintenance dollars.

Using Equation (2.21) for $i = 10\%$, $m = 7$ years, and $FW =$ \$80,000, we get

$$PW = \frac{80{,}000}{(1+0.1)^7}$$
$$= \$41{,}052.65$$

Thus, the present value of the total maintenance cost is \$41,052.65.

2.5.3 Formula for Finding Future Worth (Uniform Periodic Payments)

This formula is used to determine future worth at the end of m interest periods (in this formula an interest period is assumed one year) of equal payments made at the end of each of those m interest periods (years). The payments are deposited at interest rate i per year, compounded annually. This formula is developed as follows:

First Year

At the end of the first year, the first payment is made and the amount will be

$$AM_1 = PY \tag{2.22}$$

where

AM_1 is the amount at the end of the first year
PY is the payment made at the end of a year

Second Year

At the end of the second year, the second payment is made, as well as, interest is earned on AM_1, the amount will be

$$AM_2 = PY + AM_1(1+i) \tag{2.23}$$

where

AM_2 is the amount at the end of the second year
i is the annual compound interest rate

Substituting Equation (2.22) into Equation (2.23) yields

$$AM_2 = PY + PY(1+i) \tag{2.24}$$

Third Year

At the end of the third year, the third payment is made, as well as, interest is earned on AM_2, the amount will be

$$AM_3 = PY + AM_2(1 + i) \tag{2.25}$$

Substituting Equation (2.24) into Equation (2.25) leads to

$$\begin{aligned} AM_3 &= PY + [PY + PY(1 + i)](1 + i) \\ &= PY + PY(1 + i) + PY(1 + i)^2 \end{aligned} \tag{2.26}$$

Fourth Year

At the end of the fourth year, the fourth payment is made, as well as, interest is earned on AM_3, the amount will be

$$AM_4 = PY + AM_3(1 + i) \tag{2.27}$$

Substituting Equation (2.26) into Equation (2.27) results in

$$\begin{aligned} AM_4 &= PY + [PY + PY(1 + i) + PY(1 + i)^2](1 + i) \\ &= PY + PY(1 + i) + PY(1 + i)^2 + PY(1 + i)^3 \end{aligned} \tag{2.28}$$

Now, it is obvious that the above expression can be generalized.

***m*th Year**

At the end of the mth year, the mth payment is made, as well as, interest is earned on AM_{m-1}, the amount will be

$$AM_m = PY + PY(1 + i) + \cdots + PY(1 + i)^{m-2} + PY(1 + i)^{m-1} \tag{2.29}$$

The above expression is a geometric series which can be summed as follows:

Multiply both sides of Equation (2.29) by $(1 + i)$ to get

$$\begin{aligned} (1 + i)AM_m = PY(1 + i) + PY(1 + i)^2 + \cdots + PY(1 + i)^{m-1} \\ + PY(1 + i)^m \end{aligned} \tag{2.30}$$

Subtracting Equation (2.29) from Equation (2.30) gives

$$(1 + i)AM_m - AM_m = PY(1 + i)^m - PY \tag{2.31}$$

Rearranging Equation (2.31) yields

$$FW = AM_m = \frac{PY[(1 + i)^m - 1]}{i} \tag{2.32}$$

where FW is the future worth.

Example 2.6

Assume that for next 10 years, a person will deposit \$5,000 at the end of each of those years. The interest rate per year will be 10% compounded yearly. Find the total sum of money at the end of the specified 10-year period.

In this example, we have $PY = \$5{,}000$, $i = 10\%$, and $m = 10$ years. Using these data in Equation (2.32) will give

$$FW = AM_{10} = \frac{5{,}000[(1+0.1)^{10} - 1]}{0.1}$$
$$= \$79{,}687.123$$

Thus, the total sum of money at the end of the 10-year period will be \$79,687.123.

2.5.4 Formula for Finding Present Worth (Uniform Periodic Payments)

This formula is concerned with determining the present value of equal payments made at the end of each of m interest periods (in this formula an interest period is assumed one year). All payments are deposited at annual interest rate i compounded annually. The present worth formula is developed below:

First Year

At the end of the first year, the first payment is made and its present worth from Equation (2.21) will be

$$PW_1 = \frac{PY}{(1+i)} \tag{2.33}$$

where

PW_1 is the present worth of the payment, PY, made at the end of the first year
i is the annual compound interest rate

Second Year

At the end of the second year, the second payment is made and its present worth from Equation (2.21) will be

$$PW_2 = \frac{PY}{(1+i)^2} \tag{2.34}$$

where PW_2 is the present worth of the payment, PY, made at the end of the second year.

Third Year

At the end of the third year, the third payment is made and its present worth from Equation (2.21) will be

$$PW_3 = \frac{PY}{(1+i)^3} \tag{2.35}$$

where PW_3 is the present worth of the payment, PY, made at the end of the third year.

From Equations (2.33), (2.34), and (2.35), it is quite clear that, similarly, we can write down a present value equation for the mth payment.

***m*th Year**

At the end of the mth year, the mth payment is made and its present worth from Equation (2.21) will be

$$PW_m = \frac{PY}{(1+i)^m} \tag{2.36}$$

From Equations (2.33)–(2.36) the present worth, PWA, of all payments is

$$\begin{aligned} PWA &= PW_1 + PW_2 + PW_3 + \cdots + PW_m \\ &= \frac{PY}{(1+i)} + \frac{PY}{(1+i)^2} + \frac{PY}{(1+i)^3} + \cdots + \frac{PY}{(1+i)^m} \end{aligned} \tag{2.37}$$

This is a geometric series and can be summed as follows:

Multiply both sides of Equation (2.37) by $1/(1+i)$ to get

$$\frac{PWA}{(1+i)} = \frac{PY}{(1+i)^2} + \frac{PY}{(1+i)^3} + \frac{PY}{(1+i)^4} + \cdots + \frac{PY}{(1+i)^{m+1}} \tag{2.38}$$

Subtracting Equation (2.37) from Equation (2.38) yields

$$\frac{PWA}{(1+i)} - PWA = \frac{PY}{(1+i)^{m+1}} - \frac{PY}{(1+i)} \tag{2.39}$$

Rearranging Equation (2.39) results in

$$PWA = PY\left[\frac{1-(1+i)^{-m}}{i}\right] \tag{2.40}$$

The correctness of the above equation can easily be checked using Equation (2.32). The present value of the right-hand side of Equation (2.32) must be the same as the right-hand side of Equation (2.40). This is demonstrated below:

$$PY\left[\frac{(1+i)^m-1}{i}\right]\frac{1}{(1+i)^m}=PY\left[\frac{1-(1+i)^{-m}}{i}\right]$$

$$PY\left[\frac{1-(1+i)^{-m}}{i}\right]=PY\left[\frac{1-(1+i)^{-m}}{i}\right] \quad (2.41)$$

Example 2.7

Use the data given in Example 2.6 to find present worth of all payments.

Thus, using the data specified in Example 2.6 in Equation (2.40) yields

$$PWA=5{,}000\left[\frac{1-(1+0.1)^{-10}}{0.1}\right]$$

$$=\$30{,}722.836$$

The present worth of all payment is \$30,722.836.

2.5.5 Formula for Finding the Value of Annuity Payments When the Future Worth of the Annuity is Known

An annuity may be described as a series of equal payments, at equal intervals. Rearranging Equation (2.32), we get the value of annuity payments expressed as

$$PY=\frac{FW(i)}{(1+i)^m-1} \quad (2.42)$$

PY is the value of annuity payments
FW is the future worth of the annuity
i is the annual compound interest rate
m is the number of interest periods (years)

Example 2.8

A company desires to purchase a facility at the end of next six years. The estimated cost of the facility is \$250,000 and the company has decided to make deposits of equal sum of money at the end of each of next six years so that the total sum accumulates to $250,000. If the annual interest rate is 12% compounded annually, calculate the amount of money to be deposited at the end of each year.

In this example, we have $FW = \$250{,}000$, $i = 12\%$, and $m = 6$ years. Substituting these data into Equation (2.42) results in

$$PY = \frac{(250{,}000)(0.12)}{[(1+0.12)^6 - 1]}$$

$$= \$30{,}806.43$$

Thus, the amount of money to be deposited at the end of each year is $30,806.43.

2.5.6 Formula for Finding the Value of Annuity Payments When the Present Worth of the Annuity is Known

Rearranging Equation (2.40), the value of annuity payments is

$$PY = \frac{(PWA)(i)}{1 - (1+i)^{-m}} \tag{2.43}$$

where

PY is the value of annuity payments
PWA is the present worth of the annuity
m is the number of interest periods (years)
i is the compound interest rate per year

Example 2.9

Assume that in Equation (2.43), we have $PWA = \$80{,}000$, $i = 10\%$, and $m = 10$ years. Calculate the value of annuity payments. Utilizing these data in Equation (2.43) yields

$$PY = \frac{(80{,}000)(0.1)}{[1 - (1+0.1)^{-10}]}$$

$$= \$13{,}019.63$$

Thus, the value of annuity payments is $13,019.63.

2.6 Depreciation Methods

The meaning of the term depreciation is decline in value. Generally, it is true that the engineering items lose value with age. There are several causes responsible for the decrease in value. These are as follows [3]:

i) Functional depreciation
ii) Technological depreciation
iii) Physical depreciation
iv) Monetary depreciation

In the case of functional depreciation, a change in demand or service expected from a product during its useful life makes it less worthy from the point of view of its owner, even though it can still carry out its original mission effectively. The technological depreciation is the result of the development of better approaches of carrying out a function which makes earlier equipment designs uneconomical. In the case of physical depreciation normal wear and tear of operations slowly reduce the equipment's capability to carry out its specified function. Finally, the monetary depreciation is the result of a change in the buying power of money.

In order to take into consideration the change in the values of the engineering products, the depreciation charges are made during the useful life of such products. When computing taxable income, the depreciation charges can be deducted as operating expense. Three methods for determining depreciation are presented below.

2.6.1 Method I

This is known as the straight-line method for determining depreciation. The method assumes the linear decrease with time the value of product. During the service life of the product equal amount of money is charged for depreciation each year. The yearly depreciation charge is given by

$$CH_{DY} = \frac{AC - SV}{SL} \tag{2.44}$$

where

AC is the acquisition cost of the property
SV is the salvage value of property at the end of its service life
SL is the service life of the property in years
CH_{DY} is the yearly depreciation charge

The book value of the property at the end of year k may be obtained from the following equation:

$$BVP_k = AC - k(CH_{DY}) \tag{2.45}$$

where

BVP_k is the book value of the property at the end of year k
k is the number of years in actual service

Substituting Equation (2.44) into Equation (2.45) yields

$$BVP_k = AC - k\left[\frac{AC - SV}{SL}\right] \tag{2.46}$$

Example 2.10

Assume that the acquisition cost and the salvage value of an engineering equipment are \$250,000 and \$20,000, respectively. If the equipment's useful life is 10 years and its annual depreciation is constant, calculate the yearly depreciation charge for the equipment in question.

Substituting the given data in Equation (2.44) yields

$$CH_{DY} = \frac{(250{,}000 - 20{,}000)}{10}$$

$$= \$23{,}000$$

Thus, the yearly depreciation charge for the equipment is \$23,000.

2.6.2 Method II

This is known as the declining-balance method for determining depreciation. Sometimes, it is also called the Matheson formula or the constant percentage method. In this method, the yearly depreciation cost is a fixed percentage of the book value at the start of the year. Even though the yearly depreciation cost is different for each year, the value of the declining-balance (or fixed-percentage) factor remains constant throughout the property's service life. The method writes off the property's cost early in its life at an accelerated rate and corresponding lower yearly charges close to the final years of service. The depreciation rate (factor), DR, is defined by

$$DR = 1 - \left(\frac{SV}{AC}\right)^{1/SL} \tag{2.47}$$

The undefined symbols in Equation (2.47) are the same as used in method I.

One should note that the method assumes that the salvage value is always positive.

The book value, BVP_k, of the property at the end of year k is

$$BVP_k = AC(1 - DR)^k \tag{2.48}$$

Substituting Equation (2.47) in Equation (2.48) yields

$$\begin{aligned} BVP_k &= AC\left[1 - \left\{1 - \left(\frac{SV}{AC}\right)^{1/SL}\right\}\right]^k \\ &= AC\left(\frac{SV}{AC}\right)^{k/SL} \end{aligned} \tag{2.49}$$

The yearly depreciation charge, YDC, is expressed by

$$YDC = [BVP_{(k-1)}]DR \tag{2.50}$$

where $BVP_{(k-1)}$ is the book value at $(k - 1)$ years. Substitution of Equation (2.47) into Equation (2.50) yields

$$YDC = [BVP_{(k-1)}]\left[1 - \left(\frac{SV}{AC}\right)^{1/SL}\right] \tag{2.51}$$

Example 2.11

A piece of equipment costs \$50,000 and its expected salvage value and useful life are \$5,000 and 15 years, respectively. Calculate the book value of the equipment at the end of year 8 using straight-line and declining-balance depreciation methods.

In this example, we have $AC = \$50{,}000$, $SV = \$5{,}000$, $SL = 15$ years, and $k = 8$ years. Substituting these data in Equations (2.46) and (2.49) yield the following results, respectively:

$$\begin{aligned} BVP_8 &= (50{,}000) - 8\left[\frac{50{,}000 - 5{,}000}{15}\right] \\ &= \$26{,}000 \end{aligned}$$

$$\begin{aligned} BVP_8 &= (50{,}000)\left[\frac{5{,}000}{50{,}000}\right]^{8/15} \\ &= \$14{,}643.223 \end{aligned}$$

Thus, the book values of the piece of equipment after 8 years of service using straight-line and declining-balance depreciation methods are \$26,000 and \$14,643.223, respectively.

2.6.3 Method III

This is known as the sum-of-the-years-digits (*SYD*) method and generates the result similar to the one obtained through the declining-balance approach. The name of the method comes from the calculation procedure used. The method provides a larger depreciation charge during the early-life years of the property than during the later years. The yearly depreciation charge [4] is given by

$$YDC = (AC - SV)\left[\frac{SL - k + 1}{1 + 2 + 3 + 4 + \cdots + SL}\right]$$

$$= (AC - SV)2(SL - k + 1)/SL(SL + 1) \qquad (2.52)$$

The book value, BVP_k, of the property at the end of year k is expressed [4] by

$$BVP_k = 2(AC - SV)\left[\frac{1 + 2 + 3 + \cdots + (SL - k)}{SL(SL + 1)}\right] + SV \qquad (2.53)$$

Example 2.12

Using the data given in Example 2.11, calculate the book value of the equipment at the end of year 8 using the *SYD* method.

Thus, substituting the given data into Equation (2.53) yields

$$BVP_8 = 2(50{,}000\text{–}5{,}000)\left[\frac{1 + 2 + 3 + \cdots + (15 - 8)}{15(15 + 1)}\right] + 5{,}000$$

$$= 2(45{,}000)\left[\frac{(1 + 2 + 3 + 4 + 5 + 6 + 7)}{(15)(16)}\right] + 5{,}000$$

$$= \$15{,}500$$

Thus, the book value of the piece of equipment at the end of year 8 is $15,500.

2.7 Present Value Determination with Given Escalation and Discount Rates

Generally, each year costs are expected to rise and this factor must be taken into consideration in relevant economic analysis. When annual cost in today's dollars, as well as, escalation and discount rates are known, the following equation can be used to calculate present value [5]:

$$PV = \frac{ANC(1+j)}{(1+i)} + \frac{ANC(1+j)^2}{(1+i)^2} + \frac{ANC(1+j)^3}{(1+i)^3} + \cdots + \frac{ANC(1+j)^m}{(1+i)^m} \quad (2.54)$$

where

PV is the present value
ANC is the annual cost in today's dollars
j is the average (compound) escalation rate per year
i is the average (compound) discount rate per year
m is the number of years

Example 2.13

Assume that an equipment is to be maintained for the next four years and the following data are known:

$ANC = \$5{,}000$
$j = 5\%$
$i = 9\%$

Calculate the present value of the total expected maintenance cost.

Substituting these data into Equation (2.54) yields

$$PV = \frac{(5{,}000)(1+0.05)}{(1+0.09)} + \frac{(5{,}000)(1+0.05)^2}{(1+0.09)^2} + \frac{(5{,}000)(1+0.05)^3}{(1+0.09)^3} + \frac{(5{,}000)(1+0.05)^4}{(1+0.09)^4}$$

$$= 5{,}000\left[\frac{1.05}{1.09} + \frac{(1.05)^2}{(1.09)^2} + \frac{(1.05)^3}{(1.09)^3} + \frac{(1.05)^4}{(1.09)^4}\right]$$

$$= \$18{,}231.25$$

Thus, the present value of the total maintenance cost is \$18,231.25.

2.8 Break-even Analysis

The break-even analysis may simply be described as a graphical or algebraic representation of the relationships among cost, revenues, and volume in a company or an organization [6]. This type of analysis is extremely helpful to a manager to conceptualize such relationships. The break-even analysis is also known as cost–volume–profit analysis. The term "break-even" may be expressed as the volume needed to reach the point where total costs and revenue are equal [7]. An expression for the break-even volume is developed as follows:

At the break-even point total revenue equals total cost, thus we write

$$R_T = C_T \tag{2.55}$$

where

R_T is the total revenue
C_T is the total cost

The total revenue is given by

$$R_T = (N) \cdot (P) \tag{2.56}$$

where

P is the cost (price) of a unit
N is the number of units

The total cost is expressed by

$$\begin{aligned} C_T &= C_f + C_v \\ &= C_f + (C_{uv}) \cdot (N) \end{aligned} \tag{2.57}$$

where

C_f is the fixed cost
C_v is the total variable cost
C_{uv} is the variable cost per unit

Substitution of Equations (2.56) and (2.57) into Equation (2.55) yields

$$(N)(P) = C_f + (C_{uv})(N) \tag{2.58}$$

Rearranging Equation (2.58) results in

$$N^* = \frac{C_f}{P - C_{uv}} \tag{2.59}$$

where N^* is the number of units for break-even.

In order to find the break-even sales dollars we multiply both sides of Equation (2.58) by P to get

$$BSD = (N^*)(P) = \frac{C_f \cdot P}{P - C_{uv}} = \frac{C_f}{1 + \left[\frac{C_{uv}}{P}\right]} \tag{2.60}$$

where BSD is the break-even sales dollars.

Example 2.14

A company manufactures a number of units of an engineering equipment annually. The annual fixed cost is \$200,000. The variable cost and the selling price per unit are \$10,000 and \$20,000, respectively. Determine the number of units to break-even.

Substituting the above data in Equation (2.58) yields

$$N^* = \frac{200{,}000}{20{,}000 - 10{,}000} = 20 \text{ units}$$

In order to break-even, the company must produce 20 units of the engineering equipment.

2.9 Summary

This chapter is concerned with life cycle costing economics. The early history of interest is briefly described. Discussion on simple and compound interest is presented. A formula to calculate the total amount of money at the end of mth interst period is developed. Nominal and effective interest rates are defined. Formulas for finding future worth (single payment), future worth (uniform periodic payments), present worth (single payment), and present worth (uniform periodic payments) are presented. In addition, formulas for finding the value of annuity payments when the future worth of the annuity is known and the value of annuity payments when the present worth of the annuity is known, are developed. Three depreciation methods are described along with solved numerical examples. These methods are the straight-line, declining-balance, and sum-of-the-years-digits.

A formula to determine present value with specified escalation and discount rates is presented along with break-even analysis. Fourteen examples are solved throughout the chapter.

2.10 Problems

1. Write an essay on the history of interest.
2. Discuss the following:

   i) Simple interest
   ii) Compound interest

3. Define nominal and effective interest rates.
4. Prove

$$FW = (PA)(1 + i)^m \tag{2.61}$$

where

FW is the future worth
PA is the principal
m is the number of interest periods
i is the compound interest rate per period

5. Assume that the total maintenance cost of an equipment at the end of its 5 years service will be \$150,000. The estimated yearly compound interest rate is 15%. Calculate the present value of the maintenance dollars.
6. Prove

$$FW = PA\left[\frac{(1+j)^n - 1}{j}\right] \tag{2.62}$$

where

FW is the future worth
PA is the payment made at the end of each year
n is the number of years
j is the annual interest rate

7. Prove

$$PA = \frac{FW(j)}{(1+j)^k - 1} \tag{2.63}$$

where

PA is the value of annuity payments
j is the annual interest rate
k is the number of years
FW is the future worth of the annuity

8. Assume that the procurement cost and the salvage value of a piece of property are $50,000 and $3,000, respectively. If the property's useful life is 7 years and its annual depreciation is constant, calculate the book value of the property at the end of year 4.
9. A company produces a number of units of a product each year. The yearly fixed cost is $2 million. The variable cost and the selling price per unit are $50,000 and $90,000, respectively. Find the break-even sales dollars.

References

1. E. Paul-DeGarmo, J.R. Canada, W.G. Sullivan, Engineering Economy, Macmillan Publishing Co., Inc., New York, 1979.
2. J.E. Howell, D. Teichroew, Mathematical Analysis for Buisness Decisions, Richard D. Irwin, Inc., Homewood, Illinois, 1971.
3. J.L. Riggs, Production Systems: Planning, Analysis, and Control, John Wiley & Sons, New York, 1981.
4. J.L. Riggs, Economic Decision Models for Engineers and Managers, McGraw-Hill Book Company, New York, 1968.
5. R.J. Brown, R.R. Yanuck, Life Cycle Costing: A Practical Guide for Energy Managers, The Fairmont Press, Inc., P.O. Box 14227, Atlanta, Georgia, 1980.
6. E.E. Adam, R.J. Ebert, Production and Operations Management, Prentice-Hall, Inc., Englewood Cliffs, New Jersey 07632, 1982.

CHAPTER 3

Life Cycle Costing Basics

3.1 Introduction

An industrial product procured at the lowest initial cost may not necessarily be the one which also costs the least sum of money in the long run. The product ownership cost could be quite significant and often exceeds the acquisition cost. For example, some studies conducted in the United States military set-up indicated that the maintenance cost over the product life cycle could be many times the procurement cost. Even from the overall annual budget of the United States Department of Defense one can get the idea that the operation and maintenance cost is important. For example, in fiscal year 1974, 27% of the total Department of Defense budget accounted for operation and maintenance cost as against 20% for procurement [1]. The cost of military personnel, 28% of the Department of Defense budget, was separate. Thus, in any acquisition analysis one should also take into consideration the cost of ownership. Otherwise, the decisions based entirely on acquisition costs may not turn out to be the best decisions in the long term. The technique used to estimate the total life cycle cost of a procurement is called life cycle costing. Alternatively, life cycle costing may simply be described as a procurement process which considers the overall total cost (i.e., acquisition cost plus life cycle ownership cost) of an item. Examples of the items on which life cycle cost analysis could be quite useful are aircrafts, computers, military systems, heavy industrial equipment, automobiles, ships, appliances, hospital facilities, and buildings. This chapter discusses several introductory aspects of life cycle costing.

3.2 Need for Life Cycle Costing and Applications of Life Cycle Cost Analyses

The application of life cycle costing is following an increasing trend in recent years. There could be several factors for this upward trend. Some of those are listed below:

i) Rising inflation
ii) Budget limitations
iii) Increasing cost effectiveness awareness among users
iv) Competition
v) Costly products (e.g. commercial aircrafts, military systems)
vi) Increasing maintenance cost

Life cycle cost analyses are important in establishing, reducing, and controlling cost and have many applications. Some of those are as follows [2]:

i) Choosing the most beneficial procurement strategy
ii) Determining cost drivers
iii) Making strategic decisions and design trade-offs
iv) Selecting among options
v) Making source selections
vi) Assessing new technology application
vii) Providing objectives for program control
viii) Formulating contractor incentives
ix) Improving comprehension of basic design associated parameters in product design and development
x) Optimizing training needs
xi) Forecasting future budget needs

However, Reference [3] presents six primary uses of life cycle cost as shown in Figure 3.1. These are comparing competing projects, long-range planning and budgeting, selecting among competing bidders, controlling an ongoing project, comparing logistics concepts, and deciding the replacement of aging equipment. Another Reference [4] lists nine specific areas for the application of life cycle cost analysis. In summing up, one may state that there are a large number of areas in which the life cycle costing concept can be utilized.

3.3 Information Needed for Life Cycle Costing

In order to conduct life cycle costing studies, the required relevant information has to be identified and collected. In any case, before starting a life cycle costing study one should seek answers to questions on items such as given below [3, 5, 6]:

i) Required data
ii) Goal of the estimate

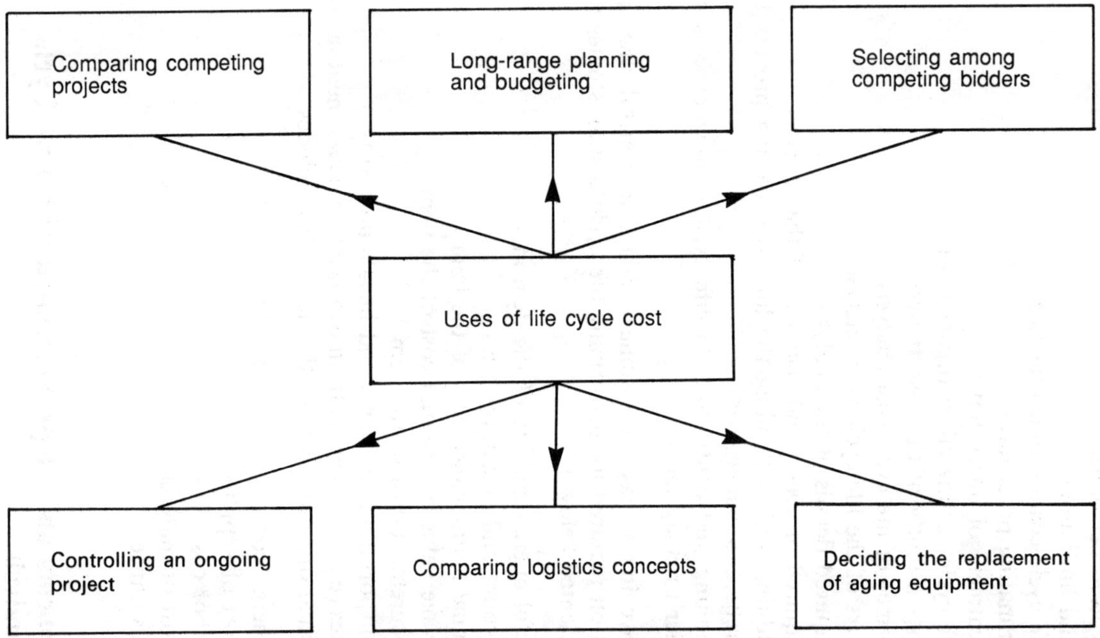

Figure 3.1. Basic uses of life cycle cost.

iii) Ground rules and assumptions
iv) Constraints associated with analysis
v) Involved personnel
vi) Fund limitations
vii) Life cycle costing time schedule
viii) Estimating procedures
ix) Treatment of uncertainties
x) Life cycle costing effort organization
xi) Responsibility of the cost analyst
xii) Format of life cycle cost analysis
xiii) Users of the life cycle cost analysis
xiv) Required details of the analysis
xv) Required precision and accuracy of the analysis
xvi) Auditing and controlling the life cycle costing process by the purchaser's management
xvii) Auditing and controlling the life cycle costing process by the seller's management

The above items may have some overlap. In general, the specific information required for performing life cycle costing studies is such as is presented below [7]:

i) Useful operational life of the item in years
ii) Discount and escalation rates
iii) Annual maintenance cost of the item
iv) Salvage value or disposal cost of the item
v) Procurement cost of the item
vi) Transportation (delivery) and installation costs
vii) Taxes (e.g., tax benefits from depreciation, investment tax credit)
viii) Annual operating cost of the item. This includes

- energy cost
- cost of supplies
- labor cost
- cost of materials
- insurance

3.4 Elements and Steps Associated with Life Cycle Cost Analysis

Before we present the steps associated with life cycle cost analysis let us consider its elements. The life cycle cost analysis includes several activities. Some of those are presented below [8]:

i) Identifying cost drivers
ii) Developing for every component in the life cycle cost breakdown structure the cost estimating relationships
iii) Developing escalated and discounted life cycle costs
iv) Defining an item or product's life cycle
v) Defining activities that generate product's ownership costs
vi) Performing sensitivity analyses
vii) Establishing constant dollar cost profiles
viii) Determining cause and effect relationships
ix) Establishing an accounting breakdown structure

Several authors have presented steps for conducting life cycle costing. The life cycle cost estimating requires numerous facts and figures. The steps proposed by three different authors are given below, separately. According to References [9, 10] the formulation of a life cycle cost can be carried out in eight steps. The steps are as follows:

i) Develop the item or equipment operating profile. This profile portrays the periodic cycle associated with the item or equipment.
ii) Develop the utilization factors. These factors indicate how or in what way the item or equipment will be operating in each operating profile mode.
iii) Identify all involved cost components
iv) Determine crucial cost related parameters
v) Compute all involved costs at present prices
vi) Escalate present costs associated with materials and labor.
vii) Discount all involved costs to base time period
viii) total all involved costs, i.e., discounted and undiscounted

According to Reference [11], the life cycle costing is composed of steps given below:

i) Determine the useful operational life of the item or product
ii) Obtain estimates for involved costs including costs of operation and maintenance
iii) Determine the item or product's terminal value
iv) Take away the terminal value from the ownership cost of the item or product
v) Discount the final amount of step (iv) to present value
vi) Add the procurement cost to the amount estimated in step (v) to obtain life cycle cost of the item or product
vii) Repeat the above steps for every item or product under consideration for procurement

viii) Make comparisons of life cycle costs of items or products being considered for purchase
ix) Purchase the item or product with least life cycle cost

A general approach to parametric life cycle cost analysis is discussed in Reference [12]. This approach is based on only five steps in contrast to earlier eight and nine steps, respectively. The five steps are shown in Figure 3.2. These are develop representative cost structure, collect data to support the cost structure, develop cost estimating

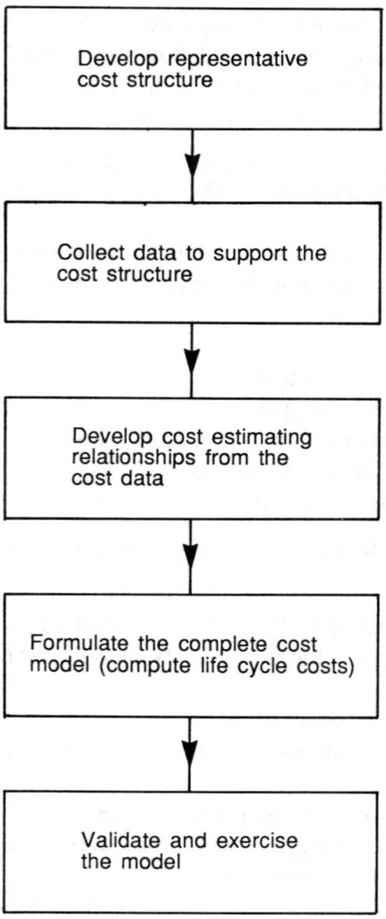

Figure 3.2. Steps associated with the general approach to parametric life cycle cost analysis.

relationships from the cost data, formulate the complete cost model (compute life cycle costs), and validate and exercise the model. This approach was developed for weapon systems and can be used to assess future programs in situations; particularly when the definitive cost data availability is limited by the modern technology. Some of the advantages of this method are that it can be employed rapidly and inexpensively, as well as, its less susceptibility to motivational bias. However, its major drawback is that it can not be applied to radically new systems.

3.5 Disciplines for Life Cycle Costing Analysts to Require Skills in and Comparative Tender Assessment Techniques

In order to perform effective life cycle costing studies, generally the life cycle costing analyst must possess skills in several disciplines. Some of those disciplines are listed below:

i) Finance and accounting
ii) Logistics
iii) Reliability and maintainability engineering
iv) Statistical analysis and quality control
v) Engineering
vi) Contracting and manufacturing engineering
vii) Estimating

Normally, it is very rare that a life cycle cost analyst has skills in all of the required disciplines. Therefore, the analyst has to seek help, as it becomes necessary, from other specialists. Usually, the professionals such as given below provide input to life cycle costing:

i) Manufacturing engineers
ii) Reliability and maintainability engineers
iii) Design engineers
iv) Logistic engineers
v) Tooling engineers
vi) Test engineers
vii) Quality control engineers
viii) Planning engineers

There are many techniques of comparative cost assessment which can be used in life cycle cost comparative tender assessment [13]. The techniques are total cost of ownership over specified period, average yearly total cost of ownership, life cycle cost versus product effectiveness, pay-back period on reliability and maintainability investment [14], and pseudo-life cycle cost with weighted present and future costs.

In the case of total cost of ownership over specified period technique, the cost considered is the total of procurement cost, ownership cost over the useful life period, and the terminal (disposal) cost (if any) discounted to present value. The average yearly total cost of ownership method divides the total cost among years for which the total cost is being estimated. A solved example is presented in Reference [7]. This technique is useful for comparing items with different useful operational life expectancies. Discussions on remaining three techniques can be found in Reference [13].

3.6 A Procedure to Incorporate Life Cycle Cost into the Planning Process for Proposals and Contracts

Every year, increasing number of equipment acquisition contracts require contractor commitments for life cycle cost. The new life cycle cost programs require estimation of costs addressing buyer and manufacturer efforts during procurement and operations. Nowadays in source selection, particularly for military projects, the life cycle cost estimates are playing an important role. Thus, one may say that the induction of life cycle costing into the acquisition process has introduced new challenges to manufacturers.

There are many contractors which have not developed the capability to carry out effective life cycle cost analyses. A request for proposal requiring an estimate of life cycle cost presents a challenging task. The task requires expertise, preparation, time, and money. In order to satisfy the life cycle cost commitments effectively, the contractor has to plan carefully and take appropriate measures. The Reference [15] presents a seven step approach for unfamiliar contractors to prepare for life cycle cost acquisitions. The steps are presented below:

i) *Organize for life cycle costing*: This is an obvious step and basically requires proper organization and placement of life cycle cost responsibilities. The responsibilities should be assigned to a group or an individual with access to decision making management. In addition, the assigned group or individual must have enough flexibility and ability to work across firm's functional organizations effectively.

ii) *Obtain and develop background information on life cycle costing*: This involves becoming acquainted with the existing life cycle cost models, components of the life cycle cost vital to the firm's product, and the life cycle cost policy directives from customers such as military (i.e., the office of the Secretary of Defense, the three Services). The Department of Defense publications, text books, conference proceedings, meetings, seminars, etc. are the sources for obtaining background information on life cycle costing. The primary objective in this step is to have working knowledge of several life cycle cost models, as well as, familiarization with the life cycle costing process.

iii) *Carry out analysis of requirements for life cycle costing response*: This step is basically concerned with obtaining particular contract proposal information for a life cycle cost acquisition. The step involves tasks such as analyses of likely life cycle cost model components to find out about the kind of data required for a life cycle cost response, determining the computer resources required to program and work with the life cycle cost model, and analysis of the information required for management decision making.

iv) *Prepare a plan for the life cycle costing technical proposal*: This step is concerned with planning the life cycle costing response for a technical proposal. Two commonly faced difficulties in a life cycle costing technical proposal are the finalization in time of the proposed product details and the product maintenance philosophy so that the detailed life cycle costing analysis can be accomplished.

v) *Develop a plan for the life cycle costing cost proposal*: This step is considered separately from step (iv) because much of the information needed for the planning for the life cycle costing cost proposal depends upon the identification of the equipment design and the maintenance philosophy developed in the earlier step (i.e. step (iv)). Many more cost estimates are required in this step. The data used in this plan must be consistent with the technical proposal plan.

vi) *Prepare a plan for identifying and analyzing life cycle cost risk*: This step is concerned with establishing a plan for identifying risk areas and addressing techniques of analyzing such risks when the life cycle cost guarantees are committed as a part of a proposed acquisition. It is imperative that all the potential risks from life cycle cost commitments must be made aware to the management.

vii) Prepare a plan for achieving life cycle cost goals: This step calls for the contractor to develop a plan for achieving the set life cycle cost goals during the contract effort. The plan should include monitoring of progress toward set goal achievement during the program phases: development, production, and operational. It is quite important that the plan for achieving life cycle cost goals must be developed early in the program, as well as, in close collaboration with all concerned, especially those responsible for providing equipment performance and maintenance experience data.

3.7 Areas for Evaluation of the Life Cycle Costing Program

The life cycle cost program has to be evaluated periodically to keep it in good order. The equipment manufacturer and user managements should perform the evaluation task. There are several areas on which questions could be asked to determine adequacy of the life cycle costing program. Some of these areas are as follows [3, 16]:

i) Incorporation of life cycle costing requirements into design subcontracts
ii) Adequate consideration of inflation and discounting factors
iii) Broadness of cost estimating data base
iv) Compatibility of reliability, maintainability, and system safety programs with life cycle cost requirements
v) Qualifications of the life cycle costing management representative
vi) Identification of cost drivers
vii) Subcontractors' cost performance review
viii) Management review of the top ten cost drivers for economy
ix) Availability of estimating support to concerned design groups and its promptness
x) Soundness of cost-estimating methods used
xi) Validation of cost estimates by an independent appraisal
xii) Performance of trade-off studies
xiii) Construction of the cost model
xiv) Coordination of life cycle cost and design to cost work
xv) Formal notification to organizations involved in the program regarding their cost goals
xvi) Awareness of the buyer concerning top ten cost drivers and suggestions for reducing such costs
xvii) Defining of cost priority relative to product delivery schedule, performance, and other requirements by the management
xviii) Awareness of planning and design groups of live cycle cost targets concerning them

3.8 Data for Life Cycle Costing

In order to have effective life cycle cost estimates of products or items, the availability of reliable historical cost data on similar items or products are vital [17]. The accuracy is often sacrificed in the event when there is lack of data sources of adequate consistency and quality, or the available data are not compatible with the estimating process. This shortcoming in life cycle costing can only be overcome by having relevant historical data.

3.8.1 Life Cycle Costing Data Base

Life cycle costing data bases are useful for performing life cycle cost studies. An organization involved in life cycle costing should have a life cycle costing data base or at least have access to it. When a new data base is being developed, the attention must be paid to factors such as follows [18]:

i) *Flexibility*: The data base should have enough flexibility to take care of local conditions effectively
ii) *Ready accessibility*: The data base should be readily accessible for retrieval, analysis, and maintenance without the help of computer support specialists
iii) *Comprehensiveness*: The data base should be comprehensive enough in scope to include all life cycle cost decision needs
iv) *Responsiveness*: The data base should be dynamic and responsive to varying needs for information
v) *Size*: The data base size should be such that it is not too large to be incomprehensible nor too small to be regarded as trivial
vi) *Expansion or contraction capability*: The data base should possess expansion or contraction capability to accommodate users' financial resources and level of participation
vii) *Orientation*: The data base should be oriented toward quantitative and qualitative factors
viii) *Uniformity*: The data base should be uniform to a degree which allows aggregation of adequate samples with similar characteristics for performing reasonable analysis

At minimum, the life cycle costing data base should include the following:

i) Cost records
ii) Procedural records: operation and maintenance

iii) User pattern records
iv) Descriptive records: hardware and site

3.8.2 Existing Data Sources

There are several publicly available information sources from which data for life cycle cost analysis can be obtained. However, the amount and quality varies with the different life cycle cost elements. For example, information on capital costs is available in considerable quantity but may not always be in the correct form. In the case of operation and maintenance costs (fuel and utilities, preventive maintenance, etc.), and cyclical renewal costs (repairs, replacements, and alterations), the sources for information are rather limited. The unit price manuals such as listed below can provide information on the construction cost component of capital costs [18].

i) Marshall & Swift
ii) Dodge
iii) Means
iv) Building Cost File
v) Richardsons

Once source for obtaining operation and maintenance cost information is the handbook published by the American Building Owners and Managers Association (BOMA). The document is published annually.

3.8.3 Areas for Questions on Data

As data are important to any life cycle costing study, there are several areas concerned with data on which questions should be asked before starting a life cycle costing study. Some of those areas are as follows:

i) Data availability
ii) Data bias
iii) Data obsolescence
iv) Data orientation towards the problem
v) Data applicability
vi) Data comparability to other existing data
vii) Data co-ordination with other information

3.9 Life Cycle Cost Report and Application of Life Cycle Costing in State Governments

The life cycle cost estimate reports are important and therefore require careful examination at the initial writing stage to see what should be in their contents. The contents of life cycle cost estimate reports may vary from one organization to another, and from one writer to another. In any case, a life cycle cost estimate report should cover the items such as the following:

i) The cost model and its details
ii) The total life cycle cost estimate expressed in dollars and essential cost breakdown estimates
iii) Cost-estimating relationships used
iv) Assumptions and ground rules
v) Data sources used and their validity
vi) The estimate's significance and conclusions
vii) Contractual cost incentives or requirements (e.g., design to cost, reliability improvements warranty, etc.), if applicable
viii) Sensitivity analyses performed
ix) The life cycle cost estimate's comparison with alternative means of estimating the same cost (if possible).

For products such as automobiles and air conditioners many state governments in the United States of America make use of the life cycle costing concept. Some of the states which make use of life cycle costing are listed below [11].

i) California
ii) North Carolina
iii) Wyoming
iv) Kentucky
v) Arkansas
vi) Washington
vii) Louisiana
viii) Connecticut

Some of the combined products on which life cycle costing is used by governments of Arkansas, California, Kentucky, Washington, and Wyoming are as follows:

i) Tractors
ii) Heat pumps
iii) Copying machines

iv) Typewriters
v) Air conditioners
vi) Refrigerators
vii) Ice machines
viii) Autos
ix) Audio-visual equipment
x) 2-way radios
xi) Medical equipment
xii) Tires
xiii) Diesel engines
xiv) Freezers
xv) Recycled motor oil
xvi) Electric ranges
xvii) Light bulbs

3.10 Benefits and Drawbacks of Life Cycle Costing and Associated Important Points

Just like any other concept, the life cycle costing too has its advantages and disadvantages. Some of the benefits of life cycle costing are: useful to control programs, an excellent tool for making a selection among the competing contractors, beneficial in comparing the cost of competing projects, useful in reducing the total cost, and making decisions associated with equipment replacement, planning and budgeting. On the other hand, some of the drawbacks are: costly, time consuming, accuracy of data is doubtful, and obtaining data for analysis are a trying task.

There are a number of important points associated with life cycle costing. Some of those are given below:

i) The management plays an important role in making the life cycle costing effort worthwhile
ii) Both the manufacturer and the user are required to organize effectively to control life cycle cost
iii) Life cycle costing is gaining importance as a technique for strategic decisions, design optimization, detailed trade-off studies, and so on
iv) The objective of life cycle costing is to obtain the maximum benefit from limited resources
v) Good data are indispensable for good life cycle cost estimates

vi) A cost analyst with excellent knowledge and experience may compensate for various data base difficulties
vii) No matter how good the estimator is some surprises may still occur
viii) The life cycle cost model must include all concerned costs associated with the program
ix) The risk management is the essence of life cycle costing
x) Throughout the life of the program, the trade-offs between life cycle cost, performance, and design to cost must be performed

3.11 Summary

This chapter presented several basic aspects of life cycle costing. Factors for the upward trend in the application of life cycle costing along with life cycle costing application areas are listed. Information required for life cycle costing is discussed. The life cycle cost analysis activities and steps for conducting life cycle costing are given. The steps proposed by three different authors are presented, separately. Disciplines for life cycle costing analysts to require skills in and comparative tender assessment techniques are outlined. An approach to incorporate life cycle cost into the planning process for proposals and contracts is described. The approach is composed of seven steps. Several areas for evaluation of the life cycle costing program are listed. The topic of "data for life cycle costing" is discussed. This covers the life cycle costing data base, existing data sources, and areas for questions on data. The contents of a typical life cycle cost estimate report are outlined and the application of life cycle costing in state governments is discussed. Some of the states which make use of life cycle costing are listed. Advantages and disadvantages of life cycle costing are presented.

3.12 Problems

1. What are the reasons for life cycle costing?
2. What information is required for performing life cycle cost analysis?
3. List at least seven components of life cycle cost analysis.
4. Describe steps for performing life cycle cost analysis.
5. What are the disciplines associated with life cycle costing?

6. What questions would you ask in evaluating a life cycle costing program?
7. What are the possible existing data sources for the life cycle costing?
8. Describe the contents of a typical life cycle cost estimate report.
9. What are the advantages and disadvantages of life cycle costing?

References

1. E. Louis-Wienecke, E.E. Feltus, Predictive Operations and Maintenance Cost Model, Report No. ADA078052, 1979. Available from the National Technical Information Service (NTIS), Springfield, Virginia, USA.
2. W.E. Lamar, Technical Evaluation Report on Design to Cost and Life Cycle Cost, North Atlantic Treaty Organization Advisory Group for Aerospace Research and Development (AGARD) Advisory Report No. 165, May 1981. Available from the National Technical Information Service (NTIS), Springfield, Virginia, USA.
3. M. Robert-Seldon, Life Cycle Costing: A Better Method of Government Procurement, Westview Press, Boulder, Colorado, 1979.
4. B.S. Blanchard, Life Cycle Costing—A Review, Terotechnica, Vol. 1, 1979, pp. 9–15.
5. H. Reiche, Life Cycle Cost, in Reliability and Maintainability of Electronic Systems, edited by J.E. Arsenault and J.A. Roberts, Computer Science Press, Potomac, Maryland 20854, 1980, pp. 3–23.
6. B.S. Dhillon, H. Reiche, Reliability and Maintainability Management, Van Nostrand Reinhold Company, New York, 1985.
7. R.J. Brown, A New Marketing Tool: Life-Cycle Costing, Industrial Marketing Management, Vol. 8, 1979, pp. 109–113.
8. M. Eddins-Earles, Factors, Formulas, and Structures for Life Cycle Costing, Published by Eddins-Earles, 89 Lee Drive, Concord, Massachusetts, 1981.
9. R.J. Kaufman, Life Cycle Costing: Decision Making Tool for Capital Equipment Acquisitions, Journal of Purchasing, Vol. 5, 1969, pp. 16–31.
10. R.J. Kaufman, Life Cycle Costing: For Capital Equipment Decisions, Automation, March 1969, pp. 75–80.
11. C.K. Coe, Life Cycle Costing by State Governments, Public Administration Review, September/October 1981, pp. 564–569.
12. H.W. Wynholds, J.P. Skratt, Weapon System Parametric Life Cycle Cost Analysis, Proceedings of the Annual Reliability and Maintainability Symposium, 1977, pp. 303–309.
13. H.S. Blanks, Traffic Signal Equipment Procurement Based on Cost of Ownership, and Community Cost, Quality and Reliability Management, Vol. 1, 1985, pp. 195–207.

14. A. Coppola, Reliability as a Capital Investment, Proceedings of the Annual Reliability and Maintainability Symposium, 1974, pp. 349–357.
15. B.A. Schmidt, Preparation for LCC Proposals and Contracts, Proceedings of the Annual Reliability and Maintainability Symposium, 1979, pp. 62–66.
16. R.L. Bidwell, Checklist for Evaluating LCC Program Effectiveness, Product Engineering Services Office, United States Department of Defense, Washington, D.C., 1977.
17. H.S. Baras, The Use of Life Cycle Cost Estimates in Weapon System Acquisitions, Industrialization Forum, Vol. 6, 1975, pp. 33–36.
18. B. Bowen, J. Williams, Life Cycle Costing and the Problems of Data, Industrialization Forum, Vol. 6, 1975, pp. 21–24.

CHAPTER 4

Life Cycle Cost Models

4.1 Introduction

There are several life cycle cost models available in published literature [1, 2]. They include both general and specific models. There is no single life cycle cost model which has been accepted as a standard model and is being used widely. There could be several factors for not having a standard one: nature of the problem, existence of many different cost data collection systems, many different types of equipment, devices or systems, the inclination of the user, etc. The factors such as these may also raise concerns for the feasibility of having a standard life cycle cost model. Nevertheless, irrespective of the types of models used for life cycle cost analysis, they all (i.e. life cycle cost models) must be visible, transparent, and effective in representing systems, equipment, subsystems, or devices.

One may describe life cycle cost models as predictive in nature and characterize a stochastic process involving many parameters. Two examples of such parameters are interest rates and reliability. In addition, costs and involved parameters (e.g., reliability and maintainability) may not be independent variables [3]. The topic of life cycle cost models is discussed in References [3, 4]. This chapter presents several selective life cycle costs models. The overall objective of all these models is to estimate life cycle cost.

4.2 Types of Life Cycle Cost Models

Some authors have classified life cycle cost models into a variety of categories. For example, Reference [5] presents six categories (as reported in Reference [3]) whereas References [3, 6] present only three. These are as follows:

i) Heuristic models
ii) Conceptual models
iii) Analytical models

A variety of analytical models can be found in published literature and the models, normally, are based on some sort of mathematical relationships. The subcategories of the analytical models cited in the literature, are the total cost models, design trade models, and logistic support models. The heuristic models may simply be described as the ill structured version of the analytical models. One example of the heuristic models can be found in Reference [7]. The hypothesized relationships of variables given in a qualitative fashion are usually the basis for the conceptual models. Normally, the conceptual models are quite flexible but have rather limited applications. One example of such models can be found in Reference [8].

In this chapter life cycle cost models are divided into two major categories: i) general life cycle cost models; ii) specific life cycle cost models.

4.3 Life Cycle Cost Model Inputs

There are a number of inputs to a life cycle cost model. They may range from training costs to shipment forecasts over the useful life of the equipment. Some of the inputs could be as follows [9]:

i) Costs of training
ii) Values of mean time between failures
iii) Mean time to repair
iv) System's listed price
v) Cost of labor per corrective maintenance action
vi) Warranty coverage period
vii) Time spent for travel
viii) Cost of installation
ix) Cost of labor per preventive maintenance action
x) Spares' requirements
xi) Average material cost of a failure
xii) Cost associated with carrying spares in inventory
xiii) Shipment forecasts over the useful life period of the equipment

4.4 General Life Cycle Cost Models

There are several life cycle cost models available in published literature.

Those models may be classified into two broad categories: specific life cycle cost models, non-specific life cycle cost models. The specific life cycle cost models are those developed for particular equipment or systems. The non-specific ones are not tied to any specific equipment or system. In this section the non-specific life cycle cost models are called "general life cycle cost models". A number of those are presented below. One should note that the models presented here may not be very general.

4.4.1 Life Cycle Cost Model I

This model was devoloped by the United States Navy for major weapon systems [10, 11]. The life cycle cost is made up of five major components: research and development cost (*RDC*), operating and support cost (*OSC*), associated systems cost (*ASC*), investment cost (*IC*), and termination cost (*TC*). Mathematically the life cycle cost is expressed as follows:

$$L_{cc} = RDC + OSC + ASC + IC + TC \tag{4.1}$$

where L_{cc} is the life cycle cost. The components of the research and development cost are validation cost and full scale development cost. The following elements belong to operating and support cost:

i) Depot supply cost
ii) Operating cost
iii) Personnel support and training costs
iv) Depot maintenance cost
v) Second destination transportation cost
vi) Organizational/intermediate maintenance activity cost
vii) Sustaining investment cost
viii) Installation support cost

The cost of associated systems has two major components: investment cost of associated systems; operating and support cost of associated systems. The investment cost is given by

$$IC = PC + GIC \tag{4.2}$$

where

IC is the investment cost
GIC is the government investment cost
PC is the procurement cost

The termination cost, TC, is defined by

$$TC = \sum_{j=1}^{N} [S(j)]\,(STC) \tag{4.3}$$

where

N is the years in life cycle
STC is the terminal cost of the major system
$S(j)$ is the number of major systems put out of action during year j

4.4.2 Life Cycle Cost Model II

In this case, the life cycle cost has two major components: recurring costs, non-recurring costs. Thus, the total life cycle cost, L_{cc}, is expressed by [12]:

$$L_{cc} = C_1 + C_2 \tag{4.4}$$

where

L_{cc} is the life cycle cost
C_1 is the recurring cost
C_2 is the non-recurring cost

The components of the recurring cost, C_1, are as follows:

i) Cost associated with maintenance
ii) Cost associated with manpower
iii) Operating cost
iv) Inventory cost
v) Support cost

Similarly, the non-recurring cost, C_2, is made up of following components:

i) Cost of training
ii) Cost of research and development
iii) Procurement cost
iv) Cost of improving reliability/maintainability
v) Support cost
vi) Qualification approval cost
vii) Cost of installation

viii) Transportation cost
ix) Cost of test equipment
x) Cost of life cycle management

4.4.3 Life Cycle Cost Model III

In this case the total life cycle cost has three major components: procurement cost, initial logistic cost, and recurring cost. The total life cycle cost is given [13] by

$$L_{cc} = PC + ILC + RC \tag{4.5}$$

where

L_{cc} is the life cycle cost
PC is the procurement cost
ILC is the initial logistic cost
RC is the recurring cost

The procurement cost represents the total of the unit prices. The one-time costs such as procurement of new support equipment not accounted in the life cycle costing solicitation and training, existing support equipment modifications, and initial technical data management belong to the initial logistic cost. The elements of the recurring cost are as follows:

i) Maintenance cost
ii) Operating cost
iii) Management cost

4.4.4 Life Cycle Cost Model IV

In this case, the life cycle cost, L_{cc}, is expressed [14, 15] by

$$L_{cc} = \sum_{j=1}^{M} CAM(k_j) \tag{4.6}$$

where

M is the quantity of modules in design k
k_j is the jth module of design k
$CAM(k_j)$ is the procurement and maintenance cost of module j generated by design k

The procurement cost in terms of candidate modules excluding spares is expressed as given below:

$$PC_{ES} = n_E \sum_{j=1}^{M} CA(k_j) \tag{4.7}$$

where

PC_{ES} is the procurement cost of quantity of n_E equipment to be procured.
$CA(k_j)$ is the procurement cost of module j of design k

Similarly, the life time support cost of n_E equipment is given by

$$C_{LS} = (POL)(M)(CMLI) + (CILI)M + n_E \sum_{j=1}^{M} CA(k_j) \cdot NSM_j \tag{4.8}$$

where

C_{LS} is the life time total organization support cost for n_E equipment
NSM_j is the module j spares to be acquired to support each equipment
$CMLI$ is the annual cost associated with maintaining a line item in inventory
POL is the equipment planned operational life expressed in years
$CILI$ is the cost associated with inducting a line item into the inventory system.

The above equation is subject to the following assumptions:

i) Each equipment is at a separate site
ii) At least one of each module will be spared at each site
iii) Discard at failure maintenance policy is followed

Adding Equations (4.7) and (4.8) results in

$$L_{cc} = n_E \sum_{j=1}^{M} CA(k_j)[1 + NSM_j] + (POL)(M)(CMLI) + (CILI)M \tag{4.9}$$

where L_{cc} is the life cycle cost. From Equation (4.9) the single equipment life cycle cost is

$$L_{ccs} = \frac{M}{n_E}[(POL)(CMLI) + (CILI)] + \sum_{j=1}^{M} CA(k_j)[1 + NSM_j] \tag{4.10}$$

where L_{ccs} is the life cycle cost of a single equipment. The above equation assumes the following [14]:

i) The cost of maintenance personnel and consumables is ignored
ii) Self-sufficiency in logistics

iii) One level of modularization
iv) The modularization of a design affects the procurement cost and the organization support cost of the physical inventory of parts.

This life cycle cost model is studied in detail in Reference [14].

4.4.5 Life Cycle Cost Model V

This model was developed by the Material Command of the United States Army [16–18]. The life cycle cost is made up of three major components: investment cost (*IC*), research and development cost (*RDC*), operating and support cost (*OSC*). Thus, the life cycle cost is expressed by

$$L_{cc} = RDC + IC + OSC \tag{4.11}$$

where L_{cc} is the life cycle cost. The research and development cost has ten cost components. These are cost of research and development data, cost of research and development facilities, cost of research and development tooling, cost of development engineering, research and development test and evaluation cost, cost of prototype manufacturing, research and development system/project management cost, research and development training services and equipment cost, cost of producibility engineering and planning, and other research and development costs.

The investment cost (*IC*) is composed of following eleven elements:

i) Cost of transportation
ii) Cost of production
iii) Cost of data
iv) Cost of initial training
v) Non-recurring investment cost
vi) Cost of initial spares and repair parts
vii) Cost of system test and evaluation
viii) Operational/site activation cost
ix) Cost of engineering changes
x) Production phase system/project management cost
xi) Other investment costs

The major components of the operating and support cost (*OSC*) are as follows:

i) Cost of indirect support operations, *CISO*
ii) Consumption cost, *CC*

iii) Cost of depot maintenance, CDM
iv) Cost of military personnel, CMP
v) Cost of material modifications, CMM
vi) Cost of other direct support operations, $CODSO$

Thus, the total operating and support cost (OSC) is given by

$$OSC = CISO + CC + CDM + CMP + CMM + CODSO \quad (4.12)$$

Equations for estimating cost elements of research and development cost, investment cost, and operating and support cost are given in Reference [11].

4.4.6 Life Cycle Cost Model VI

This model is concerned with estimating life cycle cost of systems with no redundancy. The life cycle cost is defined [19] by

$$L_{cc} = FLC + OPC + ITLC \quad (4.13)$$

where

L_{cc} is the life cycle cost
FLC is the failure cost
OPC is the operating cost
$ITLC$ is the initial cost

The initial cost is made up of all one time capital investment: initial spares and training, equipment, software, etc. The cost components such as expendables, salaries, and preventive maintenance belong to operating cost. The failure cost includes costs of repairs and operational losses attributable to system failures. The cost of an average failure can be estimated from the following expression:

$$CAF = PC + NRC + SOC + RC \quad (4.14)$$

where

CAF is the cost of an average failure
PC is the cost of parts
NRC is the non-redundancy cost
SOC is the cost of stock-out
RC is the repair cost

The non-redundancy cost is the losses which incurred by not incorporating redundancy in the equipment because of deliberate design decisions. The cost of an average failure for a simple system without any redundancy is

$$CAF = PC + CMAL + (MTTR_{AR} + T_R)CLS + CAL + (MTTR_{AR} + T_R)CPH \quad (4.15)$$

where

CAF is the cost of an average failure
PC is the replaced parts' cost
$CMAL$ is the mission-abort loss cost associated with each cessation of service. This cost is also known as the event-related, one time cost.
$MTTR_{AR}$ is the mean time to repair expressed in hours
T_R is the repair system's response time expressed in hours
CLS is the hourly cost of loss-of-service
CAL is the event-related administrative loss cost associated with each failure
CPH is the hourly cost of the purchaser's repair organization

The failure cost, *FCL*, is defined by

$$FLC = \frac{(CAF)(SLT)}{MTBF_s} \quad (4.16)$$

where

CAF is the cost of an average failure
SLT is the system's life in hours
$MTBF_s$ is the system's mean time between failures expressed in hours

Substituting Equation (4.16) into Equation (4.13) yields

$$L_{cc} = OPC + ITLC + \frac{(CAF)(SLT)}{MTBF_s} \quad (4.17)$$

Letting $HFL = CAF/MTBF_s$ in Equation (4.17) yields

$$L_{cc} = OPC + ITLC + (HFL)(SLT) \quad (4.18)$$

where HFL is the failure loss cost rate expressed in \$/hour. The cases of equipment with redundancy are studied in Reference [19].

4.4.7 Life Cycle Cost Model VII

In this case, the life cycle cost has four major components: research and development cost, production and construction cost, operation and support cost, and retirement and disposal cost. Mathematically, the life cycle cost, L_{cc}, is given [20] by

$$L_{cc} = RDC + PCC + OSC + RADC \tag{4.19}$$

where

RDC is the research and development cost
PCC is the production and construction cost
OSC is the operation and support cost
$RADC$ is the retirement and disposal cost

The research and development cost, RDC, can be estimated from the following relationship:

$$RDC = \sum_{i=1}^{7} RDC_i \tag{4.20}$$

where

RDC_i is the ith cost component of the research and development cost:

$i = 1$ (product planning)
$i = 2$ (engineering design)
$i = 3$ (system test and evaluation)
$i = 4$ (system/product life cycle management)
$i = 5$ (system/product software)
$i = 6$ (product research)
$i = 7$ (design documentation).

The production and construction cost is defined by

$$PCC = \sum_{i=1}^{5} PCC_i \tag{4.21}$$

where

PCC_i is the ith cost component of the production and construction cost:

$i = 1$ (manufacturing)
$i = 2$ (quality control)
$i = 3$ (construction)
$i = 4$ (industrial engineering and operations analysis)
$i = 5$ (initial logistics support).

The operation and support cost expression is

$$OSC = \sum_{i=1}^{3} OSC_i \tag{4.22}$$

where

OSC_i is the ith cost component of the operation and support cost:

$i = 1$ (system/product distribution)
$i = 2$ (sustaining logistic support)
$i = 3$ (system/product operations).

The retirement and disposal cost is given by

$$RADC = SURC + [\alpha(UMA)(IDC - RV)] \tag{4.23}$$

where

SURC is the system/product ultimate retirement cost
RV is the reclamation value
IDC is the cost of item disposal
α is the factor for condemnation
UMA is the number of unscheduled maintenance actions

4.4.8 Life Cycle Cost Model VIII

In the case too, the life cycle cost has four major components as follows:

i) Cost of the conceptual phase (*CCP*)
ii) Cost of the definition phase (*CDP*)
iii) Cost of the procurement phase (*CPP*)
iv) Cost of the operational phase (*COP*)

Thus, the life cycle cost is given [21] by

$$L_{cc} = CCP + CDP + CPP + COP \tag{4.24}$$

In comparison to costs of procurement and operational phases, the conceptual and definition phases costs are relatively small. These two costs are essentially associated with the labor effort. The cost elements of the procurement phase cost are as follows:

i) Prime system
ii) Personnel (acquisition)
iii) Support equipment
iv) Program management

The cost of the operational phase is given by

$$COP = MC + FOC + OAC \tag{4.25}$$

where

FOC is the functional operating cost
MC is the maintenance cost
OAC is the operational administrative cost

The above cost elements are discussed in detail in References [21–22].

4.4.9 Life Cycle Cost Model IX

This is not a life cycle cost model but has wide application in aircraft related life cycle costing. This is the reason for its inclusion in this section. The model is concerned with estimating logistics support cost and was developed by the United States Air Force [23]. The logistics support cost is made up of fuel cost, support equipment cost, software support cost, on-equipment maintenance cost, management and technical data cost, cost of pipeline and replacement spares, cost of facilities, off-equipment maintenance cost, cost of spare engines, inventory entry and supply management cost, and personnel training and training equipment cost. Mathematically, the logistics support cost is given by

$$TLSC = \sum_{j=1}^{11} LSC_j \tag{4.26}$$

where

TLSC is the total logistics support cost
LSC_j is the *i*th cost component of the logistic support cost:

$i = 1$ (fuel)
$i = 2$ (facilities)
$i = 3$ (software support)
$i = 4$ (on-equipment maintenance)
$i = 5$ (management and technical data)
$i = 6$ (pipeline and replacement spares)
$i = 7$ (support equipment)
$i = 8$ (off-equipment maintenance)
$i = 9$ (spare engines)
$i = 10$ (inventory entry and supply management)
$i = 11$ (personnel training and training equipment).

The fuel consumption cost is given by

$$LSC_1 = (T_{EF}) \cdot (C_F)(N_{AE})(R_{FC}) \tag{4.27}$$

where

LSC_1 is the fuel consumption cost
T_{EF} is the expected number of total force flying hours during the program inventory usage time period
C_F is the cost of fuel per unit (\$/gallon)
N_{AE} is the quantity of engines per aircraft
R_{FC} is the rate of fuel consumption per engine per flying hour

The cost of facilities is defined by

$$LSC_2 = (C_{NBF})(NR_{IL}) + C_{NDF} \tag{4.28}$$

where

LSC_2 is the cost of facilities
C_{NBF} is the cost associated with new base facilities to be built for system's operation and maintenance. This cost includes utilities and is expressed in dollars per base.
NR_{IL} is the quantity of intermediate repair locations
C_{NDF} is the cost associated with new depot facilities to be built for system's maintenance. This cost also includes utilities.

Estimation equations for the remaining nine components, i.e., LSC_3, LSC_4, LSC_5, LSC_6, LSC_7, LSC_8, LSC_9, LSC_{10}, and LSC_{11}, of the logistics support cost are given in Reference [23].

4.4.10 Life Cycle Cost Model X

In this case, D. B. Tague defined life cycle cost [24] as follows:

$$L_{cc} = C_{MD} + M_{MQ} + [m\{C_{MSM} + C_{BP} + C_{cc}\}] \tag{4.29}$$

where

L_{cc} is the life cycle cost
m is the quantity of systems to be purchased
C_{MSM} is the life cycle module cost of system modules
C_{BP} is the life cycle back panel cost for system
C_{MD} is the cost of module development
C_{MQ} is the module qualification cost
C_{cc} is the life cycle card cage cost for system

The module development and qualification costs are one-time costs. The components of the life cycle back panel cost include the following:

i) Wiring cost
ii) Testing cost
iii) Cost of the back panel

The cost of card cages and associated hardware belong to the life cycle card cage cost.

4.5 Specific Life Cycle Cost Models

The life cycle cost models presented in this section are generally concerned with particular systems. A number of such models are presented. One should note that some of the models presented in this section may not be very specific.

4.5.1 Specific Life Cycle Cost Model I

This life cycle cost model is concerned with retrofitting a system of avionics for a fleet of aircrafts. The life cycle cost is given [25, 26] by

$$L_{cc} = \left[\sum_{j=1}^{k} (CSR_j + CM_j) \right] + CHP + RFC \qquad (4.30)$$

where

L_{cc} is the life cycle cost
CM_j is the cost associated with maintenance for operational year j
CHP is the hardware procurement cost
CSR_j is the cost associated with spares replacement for operational year j
k is the number of years
RFC is the retrofit cost

The cost elements such as given below may be assumed to belong to retrofit cost:

i) Inspection
ii) Installation
iii) Engineering
iv) Testing
v) Drafting

The spares cost can be estimated from the following relationship if all avionic retrofit equipment is composed of line-replaceable units [26]:

$$CSR = \sum_{j=1}^{m} F_j EOH_j A \tag{4.31}$$

$$A = \sum_{p=1}^{M} \left[\frac{\{(1 - AM_p) Q_p FL_p CL_p\}}{TBMA_p} \right] \tag{4.32}$$

where

$TBMA_p$ is the line-replacement unit (LRU) p's average operating time between maintenance actions
CL_p is the pth LRU unit cost
m is the years in the life cycle
AM_p is the maintenance action's expected fraction for which the line-replaceable unit p can be repaired
M is the quantity of line-replaceable units in the system
EOH_j is the jth year's expected operating hours
FL_p is the fraction of failures for which it will not be possible to repair the line-replaceable unit p
Q_p is the number of pth LRUs in the system
F_j is the discount factor for year j

The following equation can be used to estimate maintenance cost [26]:

$$CM = \sum_{j=1}^{m} F_j EOH_j B \tag{4.33}$$

$$B = \sum_{p=1}^{M} Z_p [OW_p \{(CL_p)(OWF_p) + (OT_p)(R_{OW})\} + B_p] \tag{4.34}$$

$$B_p = \{(MC_p)(CL_p) + (MT_p)(R_m)\} MF_p \tag{4.35}$$

$$Z_p = [(1 - AM_p) Q_p] / TBMA_p \tag{4.36}$$

where

MF_p is the fraction of removals of manufacturer repaired line-replaceable unit p
R_m is the average labor rate of manufacturer expressed in dollars per man-hours
R_{OW} is the average labor rate of fleet owner expressed in dollars per man-hours
OW_p is the fraction of removals of fleet owner repaired line-replaceable unit p
OWF_p is the material cost per maintenance action incurred to fleet owner (this cost is given as a fraction of the unit cost associated with line-replaceable unit p)

MT_p is the manufacturer average man-hours incurred to repair line-replaceable unit p
MC_p is the material cost per maintenance action incurred to manufacturer (this cost is given as a fraction of the unit cost associated with line-replaceable unit p)
OT_p is the fleet owner average man-hours incurred to repair line-replaceable unit p

The assumptions associated with this model are given in Reference [26].

4.5.2 Specific Life Cycle Cost Model II

This model is concerned with estimating life cycle cost of electric motors. The life cycle cost is expressed [27] by

$$L_{mcc} = MAC + MOC \tag{4.37}$$

where

L_{mcc} is the motor life cycle cost
MAC is the acquisition cost of the motor
MOC is the operating cost of the motor

In the above equation it is assumed that the motor maintenance cost is negligible. The present worth of the motor operating cost, MOC_j, for year j is given by

$$PW_j = MOC_j\left[\frac{1}{1+i}\right]^j \tag{4.38}$$

where

PW_j is the present worth of MOC_j
i is the interest rate, fraction

If the operational life of the motor is n years then the present worth of the motor total operating cost is

$$TPW = MOC_1\left(\frac{1}{1+i}\right) + MOC_2\left(\frac{1}{1+i}\right)^2 + MOC_3\left(\frac{1}{1+i}\right)^3 + \cdots + MOC_n\left(\frac{1}{1+i}\right)^n \tag{4.39}$$

where

TPW is the present worth of the motor total operating cost
MOC_j is the motor operating cost in year j; for $j = 1, 2, 3, \ldots, n$

Assuming that the motor operating cost MOC_j does not vary much during the motor operational life, its effect is considered by an escalation factor as follows:

$$TPW = MOC\left[\left(\frac{1+ESF}{1+i}\right)+\left(\frac{1+ESF}{1+i}\right)^2+\left(\frac{1+ESF}{1+i}\right)^3 + \cdots + \left(\frac{1+ESF}{1+i}\right)^n\right] \quad (4.40)$$

where

$MOC = MOC_1 = MOC_2 = \cdots = MOC_n$, the motor operating cost.
ESF is the escalation rate, fraction

The above Equation (4.40) simplifies to the following:

$$TPW = (MOC)P\left[\frac{1-P^n}{1-P}\right] \quad (4.41)$$

where

$$P = \left(\frac{1+ESF}{1+i}\right)$$

The annual motor operating cost in dollars can be estimated from the following relationship:

$$MOC = \frac{(THY)(hp)(0.746)(CE)}{(EF)(1{,}000)} \quad (4.42)$$

where

THY is the motor operating hours in a year
hp is the motor size (horse power)
CE is the electicity cost expressed in \$/MWh
EF is the motor efficiency, fraction
MOC is the annual motor operating cost in dollars

4.5.3 Specific Life Cycle Cost Model III

This model is concerned with estimating tank gun system life cycle cost. The life cycle cost is broken into three major components: research and development cost, investment, and operating and support cost. Mathematically, the life cycle cost is expressed by

$$L_{CCTG} = G_{RDTG} + I_{TG} + C_{OSTG} \quad (4.43)$$

where

L_{CCTG} is the life cycle cost of the tank gun system
C_{RDTG} is the research and development cost of the tank gun system
I_{TG} is the investment for tank gun system
C_{OSTG} is the operating and support cost of the tank gun system

The research and development cost, C_{RDTG}, is given by

$$C_{RDTG} = \sum_{i=1}^{10} RDC_i \tag{4.44}$$

where

RDC_i is the cost component i of the tank gun system research and development cost:

$i = 1$ (tooling)
$i = 2$ (facilities)
$i = 3$ (development engineering)
$i = 4$ (system project management)
$i = 5$ (prototype manufacturing)
$i = 6$ (system test and evaluation)
$i = 7$ (training)
$i = 8$ (producibility engineering and planning)
$i = 9$ (data)
$i = 10$ (other).

The investment is given by

$$I_{TG} = \sum_{i=1}^{11} I_i \tag{4.45}$$

where

I_i is the investment component i of the tank gun system investment:

$i = 1$ (training)
$i = 2$ (production)
$i = 3$ (data)
$i = 4$ (non-recurring investment)
$i = 5$ (system project management)
$i = 6$ (initial spares and repair parts)
$i = 7$ (engineering changes)
$i = 8$ (transportation)
$i = 9$ (system test evaluation)
$i = 10$ (operational/site activation)
$i = 11$ (other).

The tank gun system operating and support cost can be estimated from the following equation:

$$C_{OSTG} = \sum_{i=1}^{6} OSC_i \tag{4.46}$$

where

OSC_i is the cost component i of the tank gun system operating and support cost:

$i = 1$ (consumption)
$i = 2$ (modification material)
$i = 3$ (military personnel)
$i = 4$ (depot maintenance)
$i = 5$ (other direct support operations)
$i = 6$ (indirect support and operations).

The elements of consumption are oil, lubricants, replenishment spares, unit training ammunition, and missiles. The labor, material, and transportation cost elements belong to depot maintenance cost. The military personnel cost includes maintenance pay, crew pay, indirect pay, allowances, etc. Maintenance civilian labor and other directs are parts of the "other direct support operations". On the other hand the indirect support and operations cost includes personnel replacement, utilities, quarters, medical support, etc.

4.5.4 Specific Life Cycle Cost Model IV

This life cycle cost model [28] represents costs associated with the entire complement of test, measurement, and diagnostic equipment (*TMDE*) of the United States Army Communications Command. The life cycle cost is made up of following 11 cost elements:

i) Documentation cost (*DC*)
ii) Holding cost (*HC*)
iii) First destination transportation cost (*FDTC*)
iv) Introduction cost (*INC*)
v) Consumables' cost (*CC*)
vi) Disposal cost (*DISC*)
vii) Hardware cost (*HDC*)
viii) Installation cost (*INSC*)
ix) Training cost (*TC*)
x) Personnel cost (*PC*)
xi) Maintenance and calibration transportation cost (*MCTC*)

Mathematically, the life cycle cost of the *TMDE* is defined by

$$L_{CCTMDE} = DC + HC + FDTC + INC + CC + DISC + HDC + INSC + TC + PC + MCTC \quad (4.47)$$

where L_{CCTMDE} is the life cycle cost of the *TMDE*. The components of the life cycle cost can be categorized into two major groups: recurring costs, non-recurring costs. The recurring costs are as follows:

i) Holding cost
ii) Consumables' cost
iii) Maintenance and calibration transportation cost
iv) Personnel cost
v) Training cost

Similarly, the non-recurring costs are as given below:

i) Hardware cost
ii) First destination transportation cost
iii) Introduction cost
iv) Documentation cost
v) Installation cost
vi) Disposal cost

The holding cost is associated with physically keeping the *TMDE* in the supply system and is defined by

$$HC = [(C_u)(N_{TMDE})(0.23)](\text{Inflation})/(\text{Discounting}) \quad (4.48)$$

where

HC is the holding cost
C_u is the unit cost
N_{TMDE} is the number of *TMDE*

The cost of consumables can be estimated from the following equation:

$$CC = [(Q_{TMDE})(FL)(0.125)(UC_{TMDE})](\text{Inflation})/(\text{Discounting}) \quad (4.49)$$

where

CC is the cost of consumables.
UC_{TMDE} is the unit cost of *TMDE*
Q_{TMDE} is the number of *TMDE* type
FL is the failures per year

The cost of documentation is expressed by

$$DC = [(NP)(C_{PG})](\text{Inflation})/(\text{Discounting}) \tag{4.50}$$

where

DC is the cost of documentation
NP is the number of documentation pages
C_{PG} is the cost associated with a single page

The documentation cost is associated with the development of the original documentation for technical manuals concerned with *TMDE* operation, repair and calibration, and so on.

The disposal cost represents the salvage value of the *TMDE* and therefore is an asset. The disposal cost can be estimated from the following relationship:

$$DISC = [(QD_{TMDE})(UC_{TMDE})(-0.1)](\text{Inflation})/(\text{Discounting}) \tag{4.51}$$

where

$DISC$ is the disposal cost
QD_{TMDE} is the quantity of *TMDE* disposed
UC_{TMDE} is the unit cost of *TMDE*

Equations for estimating *FDTC*, *INC*, *HDC*, *INSC*, *TC*, *PC*, and *MCTC* are given in Reference [28].

4.5.5 Specific Life Cycle Cost Model V

This model is concerned with estimating life cycle cost of switching power supplies [29]. The life cycle cost, L_{ccps}, is defined by

$$\begin{aligned} L_{ccps} &= C_I + C_F \\ &= C_I + \lambda(LT)(CR + CS) \end{aligned} \tag{4.52}$$

where

C_I is the initial cost
C_F is the cost of failure
λ is the constant failure rate of the unit
LT is the expected life of the product
CR is the cost of repair
CS is the spare cost

The spare cost is defined by

$$CS = (C_{us})(M_{fs}) \tag{4.53}$$

where

C_{us} is the unit spare cost
M_{fs} is the fractional quantity of spares for each active unit

Substituting Equation (4.53) into Equation (4.52) yields

$$L_{ccps} = C_I + \lambda(LT)[CR + (C_{us})(M_{fs})] \tag{4.54}$$

4.5.6 Specific Life Cycle Cost Model VI

This model is concerned with estimating the life cycle cost of a car [30]. The car life cycle cost is defined by

$$L_{ccc} = AC_c + \sum_{i=1}^{NL} (SMC_i + OC_i + URC_i) + DC \tag{4.55}$$

where

L_{ccc} is the life cycle cost of the car
NL is the lifetime of the car in years
SMC_i is the scheduled maintenance cost (tune-up, lubrication, etc) of the car for year i; $i = 1, 2, \ldots, NL$
OC_i is the operating cost (tires, gas, oil, etc) of the car for year i; $i = 1, 2, \ldots, NL$
URC_i is the unscheduled repair cost (dependent on failure rate of the car) of the the car for year i; $i = 1, 2, \ldots, NL$
DC is the disposal plus any other cost (if any)
AC_c is the acquisition cost of the car

4.5.7 Specific Life Cycle Cost Model VII

This model is concerned with estimating life cycle cost of Inertial Systems [31]. The life cycle cost is divided into three major parts: research, development, test, and evaluation cost (C_{RDTE}); procurement cost (C_P); and operation and maintenance cost (C_{OM}). Thus, the equation of the life cycle cost model for Inertial Systems is

$$L_{CCIS} = C_{RDTE} + C_P + C_{OM} \tag{4.56}$$

where L_{CCIS} is the inertial system life cycle cost. The research, development, test, and evaluation cost of the inertial system is made up of the following:

i) Cost associated with softwares
ii) Cost associated with testing
iii) Cost associated with program management
iv) Cost associated with conceptual studies
v) Cost associated with engineering change proposals
vi) Cost associated with design engineering
vii) Cost associated with technical data
viii) Cost associated with training

Equations for estimating above eight cost components are given in Reference [31]. The procurement cost has twelve distinct components as follows:

i) Cost associated with new facilities
ii) Cost associated with spares
iii) Cost associated with support equipment acquisition
iv) Cost associated with system recurring acquisition
v) Cost associated with technical data
vi) Cost associated with initial training course
vii) Cost associated with training equipment
viii) Cost associated with production tooling and test equipment
ix) Cost associated with production program start-up
x) Cost associated with initial item management
xi) Cost associated with field engineering
xii) Cost associated with equipment installation

Reference [31] presents equations for estimating the above twelve cost elements. The operation and maintenance cost is addressed at each of the three anticipated maintenance levels: line, base and depot. The basic equation for estimating operation and maintenance cost, C_{OM}, is

$$C_{OM} = \sum_{i=1}^{3} \sum_{j=1}^{M} C_{OMij} \tag{4.57}$$

where

M is the number of years
C_{OMij} is the cost associated with operation and maintenance at the ith level of maintenance in the jth year

Expressions related to Equation (4.57) can be found in Reference [31].

4.5.8 Specific Life Cycle Cost Model VIII

This model is concerned with estimating life cycle cost of avionic systems [32]. The life cycle cost is defined by

$$L_{CCAS} = ASARC + ASORC \tag{4.58}$$

where

$ASARC$ is the avionic system acquisition related cost
$ASORC$ is the avionic system operation related cost
L_{CCAS} is the avionic system life cycle cost

The avionic system acquisition related cost is given by

$$ASARC = RDTEC + DRC + ISC + FTC + LF_M(C) \tag{4.59}$$

where

$LF_M(C)$ is the initial application subsystems production cost
$RDTEC$ is the cost associated with research, development, test, and evaluation
DRC is the depot related cost
ISC is the initial spares cost
FTC is the cost associated with field tools and test equipment for local bases

The avionic system operation related costs are defined by

$$ASORC = C_1 + C_2 + C_3 + C_4 \tag{4.60}$$

where

C_1 is the cost associated with depot repairs (labor)
C_2 is the cost associated with base replacements and repairs (labor)
C_3 is the cost associated with supply system management
C_4 is the cost associated with replacement spares

Under certain assumptions, the life cycle cost can be expressed [32] as follows:

$$L_{CCAS} = A_1 + A_2 + A_3 + A_4 + A_5 + A_6 \tag{4.61}$$

where

$$A_1 = [0.6 + (1.1)LF_M]C \tag{4.62}$$

$$A_2 = [LF_1 - FL_M]C \tag{4.63}$$

$$A_3 = (1.8)(10^3)\left[\frac{LN}{12} + \frac{7}{LN}\right]\left[\frac{H}{MTBF_s}\right]M \tag{4.64}$$

$$A_4 = [LF_2 - LF_1]C \tag{4.65}$$

$$A_5 = (3.5)(10^4)\left[\frac{H}{MTBF_s}\right]\left[\frac{M}{LN}\right] \tag{4.65a}$$

$$A_6 = (3.2)(10^5) + (1.5 \times 10^4)(LN) \tag{4.66}$$

The symbols A_1, A_2, A_3, A_4, A_5, and A_6 represent procurement, initial spares, base repair, replacement spares, depot repairs, and supply system management costs, respectively. The remaining symbols are defined below.

M is the quantity of subsystems
C is the cost associated with manufacturing 1,000 systems
$MTBF_s$ is the system mean time between failures
LN is the quantity of line-replaceable units ($LRUs$)
H is the system monthly flying hours
LF_i is the manufacturing learning curve factor i: $i = 1$ (at number of systems for installation plus initial spares), $i = 2$ (at number of systems for installation plus replacement spares and initial spares)

4.5.9 Specific Life Cycle Cost Model IX

This model is concerned with estimating life cycle module cost. The life cycle module cost, L_{CCM}, is defined [24] by

$$L_{CCM} = \sum_{j=1}^{5} CM_j - T_v \tag{4.67}$$

where

CM_j is the jth cost:

- $j = 1$ (of initial spare modules per system)
- $j = 2$ (of module pipe line assets per system)
- $j = 3$ (of replenishment spare modules per system)
- $j = 4$ [(initial) of modules used per system]
- $j = 5$ (of module repair per system).

T_v is the modules' terminal value per system.

Assuming [24] $T_v = CM_2 = CM_5 = 0$, the life cycle module cost may be expressed as follows:

$$L_{CCM} = CM_1 + CM_3 + CM_4$$
$$= \sum_{j=1}^{m} C_j M_j \left[1 + \frac{NIS_j}{M_j} + \lambda_j (1 - FMR_j) SL \right] \quad (4.68)$$

where

SL is the life of the system expressed in hours
FMR_j is the ratio of disabled modules of type j which can be put back into their operational state economically to disabled repairable modules of type j
m is the module types in a system
λ_j is the hourly failures per module of type j
C_j is the cost associated with a module of type j
NIS_j is the initial spares of type j needed per system
M_j is the quantity of modules of type j

For $FMR_j = 0$, the above equation simplifies to

$$L_{CCM} = \sum_{j=1}^{m} C_j [M_j + NIS_j + M_j \lambda_j (SL)] \quad (4.69)$$

4.5.10 Specific Life Cycle Cost Model X

The model is concerned with estimating life cycle cost of software and is known as the "Boeing C-14 Model" [33–34]. The life cycle cost is expressed as follows:

$$L_{ccs} = C_A + C_S \quad (4.70)$$

where

L_{CCS} is the life cycle cost of software
C_A is the software acquisition cost
C_S is the software support cost

The software support cost is given by

$$C_S = [(DOF)(DLC)\Sigma MMS_i](1 + F) + ASC \quad (4.71)$$

where

DOF is equal to 2.5
F is the overhead factor
DLC is the direct labor cost per man-month
ASC is the other support costs (additional)
ΣMMS_i is the required man-months for support in month i

4.5.11 Specific Life Cycle Cost Model XI

This model is concerned with estimating life cycle cost of health care facilities. The life cycle cost is made up of two major components: capital cost and operating cost. Mathematically, the life cycle cost is given [11] by

$$L_{CCHCF} = CC + OC \tag{4.72}$$

where

L_{CCHCF} is the life cycle cost of health care facilities
CC is the capital cost
OC is the operating cost

The capital cost is expressed by

$$CC = \sum_{i=1}^{8} CC_i \tag{4.73}$$

where

CC_i is the component i of the capital cost:

$i = 1$ (financing cost)
$i = 2$ (indirect cost)
$i = 3$ (denial of use cost)
$i = 4$ (land acquisition cost)
$i = 5$ (direct construction or purchase cost)
$i = 6$ (collateral equipment cost)
$i = 7$ (alteration and replacement cost)
$i = 8$ (demolition and site preparation cost).

Similarly, the operating cost is expressed by

$$OC = \sum_{i=1}^{19} OC_i \tag{4.74}$$

where

OC_i is the component i of the operating cost:

$i = 1$ (utilities and fuel cost)
$i = 2$ {equipment (furnishings) maintenance cost}
$i = 3$ {equipment (fixed equipment and specific construction) maintenance cost}
$i = 4$ (space changes cost)
$i = 5$ (structural maintenance cost)

$i = 6$ (heating system operations and maintenance cost)
$i = 7$ (exterior building cleaning cost)
$i = 8$ (inspect and rodent control cost)
$i = 9$ (exterior restoration cost)
$i = 10$ (fire protection systems maintenance cost)
$i = 11$ (internal buildings cleaning cost)
$i = 12$ (painting cost)
$i = 13$ (incinerator and trash removal cost)
$i = 14$ (air conditioning and ventilating system operations and maintenance cost)
$i = 15$ (grounds and roads maintenance cost)
$i = 16$ (elevator, escalator, and dumbwaiter operations cost)
$i = 17$ (plumbing and sewerage systems operations and maintenance cost)
$i = 18$ (special mechanical systems operations and maintenance cost)
$i = 19$ (electrical system operations and maintenance cost).

4.5.12 Specific Life Cycle Cost Model XII

This model is concerned with estimating life cycle cost of appliances. An appliance life cycle cost is expressed [35] by

$$L_{CCA} = ACA + \sum_{j=1}^{LY} CEN_j \left[\frac{FC(1+R_f)^j}{(1+i)^j} \right] \tag{4.75}$$

where

L_{CCA} is the life cycle cost of the appliance
LY is the useful life of the appliance expressed in years
i is the discount rate (%) expressed in constant dollars
ACA is the acquisition cost of the appliance expressed in dollars
CEN_j is the jth year's energy consumption expressed in million BTUs
R_f is the yearly fuel escalation rate (%) expressed in constant dollars
FC is the cost of fuel in year one expressed in constant dollars per million BTUs

For constant annual energy consumption, CEN_j, and the fuel escalation rate, R_f, over the useful life of the appliance, the Equation (4.75) simplifies to

$$L_{CCA} = ACA + (CEN)(FC) \sum_{j=1}^{LY} \left[\frac{(1+R_f)}{(1+i)} \right]^j \tag{4.76}$$

The typical life times for freezers, electric dryers, water heating, room air conditioners, ranges and ovens, refrigerators, and gas dryers are 20, 14, 10, 10, 14, 15, and 11 years, respectively.

4.5.13 Specific Life Cycle Cost Model XIII

This model is concerned with estimating life cycle cost of an Early Warning Radar [36]. The total life cycle cost is made up of three major costs: procurement cost, operation cost, and logistic support cost. Thus, the radar life cycle cost is given by

$$L_{CCR} = PC + OC + LSC \tag{4.77}$$

where

L_{CCR} is the life cycle cost of the early warning radar
PC is the procurement cost of the early warning radar
OC is the operation cost of the early warning radar
LSC is the logistic support cost of the early warning radar

The life cycle cost breakdown percentages predicted for a specific early warning radar are as follows:

i) Procurement cost: 28% (of L_{CCR})
ii) Operation cost: 12% (of L_{CCR})
iii) Logistic support cost: 60% (of L_{CCR})

The procurement cost approximate breakdown percentages are as follows:

i) Fabrication cost: 20.16%
ii) Installation and checkout cost: 3.92%
iii) Design cost: 3.36%
iv) Documentation cost: 0.56%

The total of above four costs is

$$\begin{aligned} PC &= (20.16) + (3.92) + (3.36) + (0.56) \\ &\simeq 28\% \text{ of } L_{CCR} \end{aligned}$$

The operation cost approximate breakdown percentages are as follows:

i) Cost of fuel: 0.048%
ii) Cost of personnel: 8.04%
iii) Cost of power: 3.84%

The sum of above three costs is

$$OC = (0.048) + (8.04) + (3.84)$$
$$\simeq 12\% \text{ of } L_{CCR}$$

The logistic support cost approximate breakdown percentages are as follows:

i) Cost of repair labor: 38.64%
ii) Cost of initial spares: 3.25%
iii) Age cost: 1.186%
iv) Cost of repair material: 5.52%
v) Cost of initial training: 0.365%
vi) Cost of replacement spares: 11.04%

The total of above six costs is

$$LSC = (38.64) + (3.25) + (1.186) + (5.52) + (0.365) + (11.04)$$
$$\simeq 60\% \text{ of } L_{CCR}$$

4.6 Life Cycle Costing Applied to Equipment Selection

The section presents a simple hypothetical example concerned with equipment selection.

Example 4.1

Assume that a company is considering buying an engineering system. Two manufacturers A and B are bidding to sell the system. The data given in Table 4.1 are available. Which of the two systems would the company purchase?

Manufacturer A's System: The expected cost, EC_{FA}, of failure per year is

$$EC_{FA} = (0.04)(\$10{,}000) = \$400$$

The present value, PV_{MA}, of the life cycle maintenance cost is

$$PV_{MA} = EC_{FA}\left[\frac{1-(1+i)^{-m}}{i}\right] \tag{4.78}$$

where i is the interest rate and m is the system useful operating life. The factor $[(1-(1+i)^{-m})/i]$ is known as the present worth factor for uniform series.

Table 4.1. Data for systems under consideration.

No.	Description	Manufacturer *A*'s system	Manufacturer *B*'s system
1	Selling price	\$100,000	\$120,000
2	Constant failure rate per year	0.04 failures/ year	0.05 failures/ year
3	Cost of money (interest rate per annum)	10%	10%
4	Expected operating life	10 years	10 years
5	Expected cost of a failure	\$10,000	\$12,000
6	Expected annual operating cost	\$6,500	\$3,000

In this example, we have $i = 10\%$, $EC_{FA} = \$400$, and $m = 10$ years. Thus, substituting the above data into Equation (4.78) yields

$$PV_{MA} = (400)\left[\frac{1-(1+0.1)^{-10}}{0.1}\right] = (400)(6.1446)$$

$$= \$2457.83$$

Similarly, the present value, PV_{OA}, of the life cycle operating cost is

$$PV_{OA} = (6500\left[\frac{1-(1+0.1)^{-10}}{0.1}\right]$$

$$= \$39,939.69$$

Thus, the life cycle cost, L_{CCMAS}, of manufacturer *A*'s system is

$$L_{CCMAS} = SPMAS + PV_{MA} + PV_{OV} \tag{4.79}$$

where *SPMAS* is the selling price of manufacturer *A*'s system. By Substituting data for *SPMAS*, PV_{MA}, and PV_{OA}, Equation (4.79) yields

$$L_{CCMAS} = (100,000) + (2457.83) + (39,939.69)$$

$$= \$142,397.52$$

Manufacturer B's System: The expected cost, EC_{FB}, of failure per year is

$$EC_{PB} = (0.05)(12{,}000) = \$600$$

The present value, PV_{MB}, of the life cycle maintenance cost is

$$PV_{MB} = 600\left[\frac{1-(1+0.1)^{-10}}{0.1}\right]$$
$$= \$3{,}686.74$$

Similarly, the present value, PV_{OB}, of the life cycle operating cost is

$$PV_{OB} = (3{,}000)\left[\frac{1-(1+0.1)^{-10}}{0.1}\right]$$
$$= \$18{,}433.7$$

The life cycle cost, L_{CCMBS}, of manufacturer B's system is

$$L_{CCMBS} = (120{,}000) + (3{,}686.74) + (18{,}433.7)$$
$$= \$142{,}120.44$$

It would be slightly more economical for the company to purchase manufacturer B's system because its life cycle cost is lower than that of manufacturer A's system.

4.7 Summary

This chapter presents several life cycle cost models. The types of life cycle cost models are discussed along with life cycle cost model inputs. Life cycle cost models presented are grouped into two broad categories: specific life cycle cost models and non-specific (general) life cycle cost models. The specific life cycle cost models are those developed for particular equipment or systems. The non-specific (general) ones are not tied to any specific equipment or system. Ten generally non-specific (general) life cycle cost models are presented. These include models developed by the United States Navy, Army, and Air Force. In addition, thirteen specific life cycle cost models are presented. These models deal with items such as the following:

i) Motors
ii) Tank gun systems
iii) Test, measurement, and diagnostic equipment
iv) Switching power supplies

v) Cars
vi) Inertial systems
vii) Avionic systems
viii) Modules
ix) Computer software
x) Health care facilities
xi) Appliances
xii) Early warning radars

A simple and solved hypothetical example concerned with equipment selection is presented.

4.8 Problems

1. What is a life cycle cost model?
2. What are the types of life cycle cost models?
3. What are the inputs to a life cycle cost model?

Table 4.2. Data for equipment under consideration.

No.	Description	Manufacturer *A*'s equipment	Manufacturer *B*'s equipment	Manufacturer *C*'s equipment
1	Expected cost of a failure	$10,000	$11,000	$12,000
2	Failure rate per annum	0.05 failures/year	0.06 failures/year	0.07 failures/year
3	Expected annual operating cost	$10,000	$9,000	$6,000
4	Cost of money (interest rate per annum)	15%	15%	15%
5	Expected useful operating life	15 years	15 years	15 years
6	Selling price	$90,000	$100,000	$110,000

4. Compare the two "general" life cycle cost models developed by the United States Army and Navy, separately.
5. List the characteristics of an ideal life cycle cost model.
6. Assume that an organization is considering to purchase an engineering equipment. Three manufacturers *A*, *B*, and *C* are bidding to sell the equipment. The data presented in Table 4.2 are available. Which of the three equipment would be the most economical for the organization to procure?

References

1. B.S. Dhillon, Life Cycle Cost: A Survey, Microelectronics and Reliability: An International Journal, Vol. 20, 1980, pp. 737–742.
2. B.S. Dhillon, Reliability Engineering in Systems Design and Operation, Van Nostrand Reinhold Company, New York, 1983.
3. Y.P. Gupta, Life Cycle Cost Models and Associated Uncertainties, in Electronic Systems Effectiveness and Life Cycle Costing, edited by J.K. Skwirzynski, Springer-Verlag, Berlin, 1983, pp. 535–549.
4. B. DeNeumann, Life Cycle Cost Models, in Electronic Systems Effectiveness and Life Cycle Costing, edited by J.K. Skwirzynski, Springer-Verlag, Berlin, 1983, pp. 513–532.
5. L.E. Dover, B.E. Oswald, A Summary and Analysis of Selected Life Cycle Costing Techniques and Models, Masters Thesis, 1974, Air Force Institute of Technology, Wright-Patterson Air Force Base, Ohio, USA.
6. Y.S. Sherif, W.J. Kolarik, Life Cycle Costing: Concepts and Practice, OMEGA, Vol. 9, 1981, pp. 287–296.
7. W.J. Kolarik, Analysis Theory and Procedures for Determining and Predicting Availability, Availability Cost, and Intangible Effect for Farm Machinery Systems, Ph.D. Dissertation, 1977, Oklahoma State University, Stillwater Oklahoma.
8. A.S. Goldman, T.B. Slattery, Maintainability: A Major Element of System Effectiveness, John Wiley & Sons, New York, 1967.
9. D.P. Siewiorek, R.S. Swarz, The Theory and Practice of Reliable System Design, Digital Press, Digital Equipment Corporation, Bedford, Massachussets, 1982.
10. Life Cycle Cost Guide for Major Weapon Systems, Prepared by Naval Weapons Engineering Support Activity, Naval Material Command, Pentagon, Washington, D.C., January 1977.
11. M. Eddins-Earles, Factors, Formulas, and Structures for Life Cycle Costing, Published by Eddins-Earles, 89 Lee Drive, Concord, Mass. 01742, 1981.
12. H. Reiche, Life Cycle Cost, in Reliability and Maintainability of Electronic Systems, edited by J.E. Arsenault and J.A. Roberts, Computer Science Press, Potomac, Maryland 20854, 1980, pp. 3–23.

13. M.O. Locks, Maintainability and Life-Cycle Costing, Proceedings of the Annual Reliability and Maintainability Symposium, 1978, pp. 251–253.
14. J.E. Biegel, A Multilevel Modularization Technique Designed to Minimize Life Cycle Costs for Large Systems, Proceedings of the Spring Annual Conference of the American Institute of Industrial Engineers, 1979, pp. 335–342.
15. A. J. Caponecchi, A Methodology for Obtaining Solutions to the Modular Design Problem, Ph.D. Dissertation, December 1971, The University of Texas at Austin, Austin, Texas.
16. "Investment Cost Guide for Army Material Systems", Pamphlet No. 11-3, Department of the Army, United States Department of Defense, Alexandria, VA, April 1976.
17. "Research and Development Cost Guide for Army Material Systems", Pamphlet No. 11-2, Department of the Army, United States Department of Defense, Alexandria, VA, May 1976.
18. "Operating and Support Cost Guide for Army Material Systems", Pamphlet No. 11-4, Department of the Army, United States Department of Defense, Alexandria, VA, April 1976.
19. D.B. Dickinson, L. Sessen, Life Cycle Cost Procurement of Systems and Spares, Proceedings of the Annual Reliability and Maintainability Symposium, 1976, pp. 282–286.
20. B.S. Blanchard, Design and Manage to Life Cycle Cost, M/A Press, Portland, Oregon, 1978.
21. N.C. Stordahl, J.L. Short, The Impact and Structure of Life Cycle Costing, Proceedings of the Annual Symposium on Reliability, 1968, pp. 509–515.
22. B.S. Dhillon, Reliability Engineering in Systems Design and Operation, Van Nostrand Reinhold Company, New York, 1983.
23. "Logistics Support Cost Model Users Handbook, Acquisition Logistics Division, Air Force Logistics Command, United States Air Force, Wright-Patterson Air Force Base, Ohio, January 1979.
24. A. Davies, K.J. Skinner, The Application of a Life Cycle Cost Model to Modular Electronic Systems, The Radio and Electronic Engineer, Vol. 53, 1983, pp. 209–215.
25. E. Andrew, et al, Microwave Landing System Integration Study, Report No. GSE-SE-74-1, 1974, Air Force Institute of Technology, Wright Patterson Air Force Base, Ohio.
26. T.L. Regulinski, Y.P. Gupta, Reliability Cost Estimation: Managerial Perspectives, IEEE Transactions on Reliability, Vol. 32, 1983, pp. 276–281.
27. V. Ganapathy, Life Cycle Costing Applied to Motor Selection, Process Engineering, July 1983, pp. 51–52.
28. H. Rosenberg, J.H. Witt, Effects on *LCC* of Test Equipment Standardization, Proceedings of the Annual Reliability and Maintainability Symposium, 1976, pp. 287–292.

29. D. Monteith, B. Shaw, Improved *R*, *M*, and *LCC* for Switching Power Supplies, Proceedings of the Annual Reliability and Maintainability Symposium, 1979, pp. 262–265.
30. S.K. Bhuyan, Cost of Quality as a Customer Perception, Proceedings of the American Society for Quality Control Annual Congress, 1982, pp. 459–464.
31. D.E. DeBurkarte, A Standard Life Cycle Cost Model for Inertial Systems, Proceedings of the IEEE 1976 National Aerospace and Electronics Conference (NAECON' 76), 1976, pp. 687–695.
32. P.S. Kilpatrick, A.L. Jones, Life Cycle Cost Comparisons of Avionic System Design Alternatives, Proceedings of the IEEE 1974 National Aerospace and Electronics Conference, 1974, pp. 514–520.
33. "Advanced Avionics Systems for Multi-Mission Applications", Report—Vol. II—Appendix G, Boeing Aerospace Company, Seattle, Washington, 1978.
34. D.V. Ferens, R.L. Harris, Avionics Computer Software Operation and Support Cost Estimation, Proceedings of the IEEE 1979 National Aerospace and Electronics Conference (NAECON '79), 1979, pp. 296–300.
35. I. Turiel, H. Estrada, M. Levine, Life Cycle Cost Analysis of Major Appliances, Energy, Vol. 6, 1981, pp. 945–970.

CHAPTER 5

Cost Estimation Techniques

5.1 Introduction

Cost estimating is a popular activity as various kinds of costs have to be estimated each day. The estimated cost has to be as close as possible to its actual value (if not exactly the same as the actual value), otherwise, it may lead to a consequence of a severe degree. More specifically, the success or failure of an organization may depend on the quality of cost estimates. There are over 400,000 companies that produce products of one or other type in the United States. In all of them the cost estimating activity is mandatory. In order to get some feedback regarding the patterns of organization, staffing, policy formulation, and work content of the cost estimating activity in such companies, surveys of 104 companies were conducted [1]. These surveys revealed that cost estimators belonged to different backgrounds: industrial and manufacturing engineers, design engineers, accountants, etc. In addition, the distribution of departmental responsibility for the cost estimating function was as following:

i) Manufacturing (43%)
ii) Accounting (20%)
iii) Engineering (17%)
iv) Sales and marketing (12%)
v) Others (8%)

The above diversities make the cost estimating function even more complex and demanding. There are several methods and procedures developed over the years for estimating costs. Each one of them has its strong and weak points. A cost estimation method or procedure may be very attractive for one type of application and less effective for another. Therefore, one should take a great degree of care in applying those techniques. This chapter presents some of the existing methods and models.

5.2 Cost Estimation Methods

Over the years, people working in cost estimating area have come up with several different cost estimation methods, models, and procedures. Many of those can be found in the published literature [2–8]. Some of the methods are presented below:

5.2.1 Cost Estimation Method I

In the published literature, this method is referred to as cost indexes. A cost index may simply be described as a dimensionless number used for the purpose of adjusting an item's cost from one period to another [6]. More specifically, the index allows one to estimate the cost of a similar item without going through the rigor of detailed costing. The history of the index numbers goes back to the middle of eighteenth century when G.R. Carli, an Italian citizen, made use of these numbers to examine the effects of the discovery of the New World on the purchasing power of money in the continent of Europe [7].

There are three reasons for items' cost to change continuously: inflation, changing technology, and changing availability of materials and labor. The following equation can be used for converting costs from one period to another:

$$C_k = C_m\left(\frac{I_k}{I_m}\right) \tag{5.1}$$

where

C_k is the cost, in dollars, in year k
C_m is the cost, in dollars, in year m
I_m is the value of the index for year m
I_k is the value of the index for year k

Example 5.1

A \$0.5 million engineering system was manufactured in 1981. For that year the index was 721. If the index value for 1985 is 790, calculate the cost of the system in 1985. In this example we have

$$I_{(1985)} = 790$$

$$I_{(1981)} = 721, \text{ and}$$

$$C_{(1981)} = \$0.5 \text{ million}$$

Thus, Equation (5.1) yields

$$C_{(1985)} = C_{(1981)}\left[\frac{I_{(1985)}}{I_{(1981)}}\right]$$
$$= 0.5\left[\frac{790}{721}\right]$$
$$= \$0.5479 \text{ million}$$

The cost of the system in 1985 will be $0.5479 million.

There are numerous indexes used for the purpose of estimating costs. Some of the common cost indexes are as follows:

i) Chemical Engineering Plant Cost Index [9]
ii) Marshall and Swift Equipment Cost Index [10]
iii) Engineering News-Record (ENR) Building and Construction Cost Indexes [11]
iv) Nelson Refinery Construction Cost Index [12]
v) Bureau of Labor Statistics Indexes [13–15]

Table 5.1. Client-oriented specialized indexes and foreign indexes.

No.	Client-oriented specialized indexes	Indexes being published in foreign countries
1	DRI Index (Published by Data Resources Inc.)	—
2	SRI Index (Published by Stanford Research Institute)	—
3	Chase Econometric Inflation Index	—
4	—	Plant Cost Index, Canada [17]. The base year is 1971 (100).
5	—	Chemical Plant Cost Index, U.K. [18]. The base year is 1970 (100).
6	—	Plant Cost Index, West Germany [19]. The base year is 1962 (100).

Table 5.1 lists some client-oriented specialized indexes and indexes being published in some foreign countries. We should note here that the cost indexes are subject to several limitations [16]: very limited accuracy for periods more than a decade, represent composite data, average data, etc. Each of the five common cost indexes presented earlier is discussed below.

Chemical Engineering (*CE*) Plant Cost Index

This is a complex index first introduced in February 1963. For the period 1957–1959 the index was 100. This index may simply be described as an overall index for the cost of constructing chemical process plants [20] and is composed of the following four major components [9]:

i) Labor for installation and erection
ii) Building, material, and labor
iii) Equipment, machinery, and supports
iv) Engineering and supervision manpower

The percentages associated with above four components are 22%, 7%, 61%, and 10%, respectively. The third component (equipment, machninery, and supports) is subdivided into several categories [22]: process machinery (14%); electrical equipment and materials (5%); pipes, fittings, and valves (20%); insulation, structural supports, and paint (10%); pumps and compressors (7%); assembled equipment (37%); and process instruments and controls (7%).

The *CE* index does not take into consideration the following items [22]:

i) Insurance
ii) Contractors' profits during construction
iii) Site clearing and preparation cost
iv) Overhead
v) Taxes

A survey of 156 plants/projects was conducted to derive weight factors for the index. The U.S. Bureau of Labor Statistics is the source of most of the price data used. In 1982, the cost components of the index were updated. The reason for this update was to take into consideration the changes in productivity. Table 5.2 presents rounded off values for the *CE* plant cost index for selected years [21].

Table 5.2. Selected values for the *CE* plant cost index.

Year	Value of the *CE* plant cost index (for 1957–59, the index value was 100)
1965	104
1967	110
1969	119
1971	132
1973	144
1975	182
1977	204
1979	239
1981	297
1982	314
1983	317
1984	323
1985	325

Marshall and Swift (*M&S*) Equipment Cost Index

This index (formerly Marshall and Stevens) represents the installed equipment cost average for 47 different industrial, commercial and housing industries and therefore may be described as an overall index for the cost of industrial equipment. The value of this index was 100 for 1926. It includes industries such as those listed below.

i) Paper
ii) Chemical
iii) Rubber
iv) Clay products
v) Petroleum
vi) Glass
vii) Cement

Table 5.3 presents selected rounded-off values for the *M&S* equipment cost index [21].

Engineering News-Record (*ENR*) Building and Construction Cost Indexes

The Engineering News-Record (*ENR*) Construction Cost Index was first introduced in 1921 for the purpose of tracking the industrial construction costs. Usually the *ENR* indexes are based on index value

Table 5.3. Some rounded-off values for the *M&S* equipment cost index.

Year	Value of the *M&S* equipment cost index (for 1926, the index value was 100)
1965	245
1967	263
1969	285
1971	321
1973	344
1975	444
1977	505
1979	599
1981	721
1982	746
1983	761
1984	780
1985	790

of 100 for 1913. However, a reference value of 100 for 1967 and 1949 is also being used. The components of the *ENR* construction cost index are as follows:

i) Lumber
ii) Structural steel
iii) Portland cement
iv) Common labor

Relative weights associated with above components are 9%, 15%, 2%, and 74%, respectively. Similarly, the components of the Building Cost Index are

i) Skilled labor
ii) Portland cement
iii) Lumber
iv) Structural steel

In addition, their (the components) relative weights are 56%, 3%, 15%, and 26%, respectively. Some of the points to be noted with *ENR* indexes are as follows [21]:

i) The productivity factors like the ones in Chemical Engineering (*CE*) Index are not included

ii) The seasonal adjustment in costs is not carried out
iii) Factors such as managerial efficiency, competitive conditions, and overhead of profits are not incorporated
iv) The cost differentials among different cities are not taken into consideration

Nelson Refinery Construction Cost Index

This index is very useful for petroleum industry and is based on 60% labor and 40% materials costs. The labor is subdivided into two components: common (21%) and skilled (39%). Similarly, the three components of materials are non-metallic building materials (8%), iron and steel (20%), and miscellaneous fabricated equipment (12%).

This is an inflation index which uses the index value of 100 for 1946. The productivity changes are not incorporated into this index. Table 5.4 presents selected rounded-off values for the Nelson Refinery Construction Cost Index.

Bureau of Labor Statistics (*BLS*) Indexes

The Bureau of Labor Statistics of the United States Department of Labor publishes data on the prices of hundreds of items. The consumer Price Index and inflation rates for various classifications of products are computed using these data. The average earnings statistics are also published by the Bureau of Labor Statistics. These statistics provide input to labor and material indexes. The *BLS* data are used in most of the earlier described cost indexes [21].

Table 5.4. Some selected values for the Nelson Index.

Year	Value of the Nelson cost index (for 1946, the index value was 100)
1971	141
1973	163
1975	200
1977	223
1979	264

For periodic update of cost indexes, one should consult periodicals such as the following:

i) Oil and Gas Journal (Nelson Refinery Construction Cost Index)
ii) Engineering News-Record (Engineering News-Record (*ENR*) Indexes)
iii) U.S. Bureau of Labor Statistics (Consumer Price Index, Producer Price Indexes, etc.)
iv) Chemical Engineering (Chemical Engineering (*CE*) Plant Cost Index, Marshall and Swift (*M&S*) Equipment Cost Index)

For the purpose of converting all indexes to a common base year, the author of Reference [22] presents the following relationship:

$$I_{vnb} = \frac{I_{vobc}}{I_{vobn}} \tag{5.2}$$

where

I_{vnb} is the value of the index on the new base
I_{vobc} is the value of the index (old base) to be converted
I_{vobn} is the value of the index (old base) for the new base year

5.2.2 Cost Estimation Method II

This method is often referred to as Cost-Capacity Factors. The method is used to approximate cost estimates for equipment or for plants where cost data for similar projects of different capacity than desired, are known. The cost-capacity relationship is defined by the following equation:

$$C_N = C_{OL}\left[\frac{CP_N}{CP_{OL}}\right]^{\alpha} \tag{5.3}$$

where

C_N is the estimated cost of equipment/plant of desired capacity, CP_N
C_{OL} is the known cost of similar equipment/plant of capacity, CP_{OL}
α is the cost-capacity factor. The frequently used value for α for processing equipment is near 0.6.

However, Table 5.5 presents values of α for selected equipment [6, 8, 20].

Table 5.5. Selective values for the cost-capacity factor.

Number	Equipment	Value of the cost-capacity factor (α)
1	Pumps	0.6
2	Tanks	0.7
3	Conveyor	0.7
4	Heaters	0.8
5	Blower, centrifugal (size: 10^3–10^4 ft^3/min.)	0.59
6	Fan, centrifugal (size: 10^3–10^4 ft^3/min.)	0.44
7	Heat exchangers	0.6
8	Motor: Explosion proof, 400 volts, induction, squirrel cage	0.99 (20–200 horse power) 0.69 (5–20 horse power)
9	Refrigeration unit	0.75
10	Transformer, 3-phase (size: 9–45 kw)	0.47
11	Dust collector, cyclone (size: 2–7,000 ft^3/min.)	0.61

Example 5.2

Assume that the cost of a plant with a capacity of 250,000 tons/year is \$40 million. Calculate the cost of a similar 400,000 tons/year plant, if the value of the cost-capacity factor is 0.7.

Using the given data in Equation (5.3) yields

$$C_N = 40\left[\frac{400{,}000}{250{,}000}\right]^{0.7}$$

$$= \$55.58 \text{ million}$$

Thus, the cost of a 400,000 tons/year plant will be \$55.58 million. Probably the simplest approach for estimating plant costs is to make use of cost-capacity factors along with cost indexes to take care of inflation, as follows:

$$C_N = C_{OL}\left[\frac{CP_N}{CP_{OL}}\right]^{\alpha}\left[\frac{I_k}{I_m}\right] \tag{5.4}$$

Example 5.3

Assume that the total investment cost of a processing plant with 100,000 tons/year capacity was \$15 million in 1981. How much would a 200,000 tons/year similar plant have cost in 1985? The *CE* indexes for 1981 and 1985 were 297 and 325, respectively. The value of the cost-capacity factor is 0.6.

Substituting the given data into Equation (5.4), we get

$$C_N = 15\left[\frac{200{,}000}{100{,}000}\right]^{0.6}\left[\frac{I_{(1985)}}{I_{(1981)}}\right]$$

$$= 15\left[\frac{200{,}000}{100{,}000}\right]^{0.6}\left[\frac{325}{297}\right]$$

$$= \$24.88 \text{ million}$$

Thus, in 1985 the 200,000 tons/year plant's total investment cost would have been \$24.88 million.

5.2.3 Cost Estimation Method III

This method is known as the Lang Factor Method. The technique was developed by H. J. Lang [23] to obtain quick order-of-magnitude cost estimates by using historical average cost factors. Lang proposed to estimate total plant costs from the delivered equipment cost using the following factors as multipliers:

i) $k = 4.74$ (for fluid process plants)
ii) $k = 3.10$ (for solid process plants)
iii) $k = 3.63$ (for solid–fluid plants)

Thus, the total estimated plant cost, C_{TP}, is given by

$$C_{TP} = kC_{DE} \tag{5.5}$$

where

k is the Lang factor depending on the nature of the plant
C_{DE} is the delivered equipment cost

Example 5.4

A solid-processing plant's delivered equipment cost is \$15 million. Estimate the total plant cost.

Thus, the total plant cost, *TPC*, is

$$TPC = (3.10)(15)$$
$$= \$46.5 \text{ million}$$

5.2.4 Cost Estimation Method IV

This is known as the Rudd and Watson Method [24]. The fixed capital cost is obtained from the product of delivered equipment cost and multiplying factors greater than unity. The fixed capital cost, *FCC*, is expressed by the following equation:

$$FCC = \prod_{i=1}^{n} \alpha_i \Sigma PEC \qquad (5.6)$$

where

α's are the factors for the various items (components for installation, etc.). Examples of these items are listed in Reference [6].
PEC is the purchased equipment cost

5.2.5 Cost Estimation Method V

This is known as the Wessell Method [16]. The method is useful for making an order-of-magnitude approximation of operating labor requirements in the absence of a manning table. The Wessell equation is defined below.

$$\frac{OMH}{TP} = \theta\left[\frac{PS}{(CP)^{0.76}}\right] \qquad (5.7)$$

where

CP is the capacity, tons per day
TP is the tons of product
PS is the number of process steps
OMH is the operating man-hours

The values of θ for a batch operation with maximum labor, well-instrumented continuous process operation, and operation with average labor needs are 23, 10, and 17, respectively.

5.2.6 Cost Estimation Method VI

This method is referred to as the Hand Factor Method. In 1958, W.E. Hand [25] proposed a refinement of the Lang Factor Method. In this refinement he recommended the use of the different factor for the several groups of equipment. The total installed cost for each equipment group j is expressed as follows [2]:

$$TIC = k_j C_{DEj} \tag{5.8}$$

where

TIC is the total installed cost
C_{DEj} is the delivered equipment cost
k_j is the Hand factor. The factor covers field materials (insulation, structures, electrical, piping, foundations, and finishes), indirect costs, and labor.

Hand factors for various equipment are as follows:

i) $k_j = 4$ (pumps)
ii) $k_j = 4$ (pressure vessels)
iii) $k_j = 2.5$ (miscellaneous equipment)
iv) $k_j = 3.5$ (heat exchangers)
v) $k_j = 4$ (fractionating towers)
vi) $k_j = 2$ (fired heaters)
vii) $k_j = 4$ (instruments)
viii) $k_j = 2.5$ (compressors)

An additional list of factors is presented in Reference [26]. One should note here that the factors do not incorporate a contingency allowance.

5.2.7 Cost Estimation Method VII

This is known as the Turnover Ratio Method. The method is most efficient for estimating plant costs but probably is the least accurate. The turnover ratio is expressed as follows:

$$TR = \frac{GAS}{FCI} \tag{5.9}$$

where

GAS is the gross annual sales
TR is the turnover ratio
FCI is the fixed capital investment

The gross annual sales is given by

$$GAS = (USP)(APR) \tag{5.10}$$

where

USP is the unit selling price
APR is the annual production rate

The value of the turnover ratio vary from about 0.2 to 8.0. In particular, for the chemical industry its value is roughly from 1.00 to 1.25.

Example 5.5

A manufacturing plant is to produce 100,000 units/year. Each unit's selling price is $200. Estimate the fixed capital investment, if the turnover ratio is 2.

Substituting Equation (5.10) and the given data into Equation (5.9), we get

$$FCI = \frac{(USP)(APR)}{TR}$$

$$= \frac{(200)(100{,}000)}{2}$$

$$FCI = \$10{,}000{,}000$$

Thus, the fixed capital investment for the manufacturing plant is $10 million.

5.2.8 Cost Estimation Method VIII

This is called the Factor Method and is important for detailed estimates. The following relationship is used to estimate the value (e.g. cost, price, etc.) of the design being evaluated [7]:

$$K = [CSE + A](\beta + 1) \tag{5.11}$$

where

$$A = \sum_j \alpha_j (CSE) \tag{5.12}$$

The symbols used in the above equations are defined below:

CSE is the selected major equipment's cost
α_j is the factor for the estimating of instrumentation, buildings, etc
k is the value of design
β is the factor for the estimating of indirect costs: contingency, engineering, etc

This method is described in detail in Reference [7].

5.2.9 Cost Estimation Method IX

This is known as the Ratio Factor Method. The capital investment, *CIP*, for a plant is estimated from the following relationship:

$$CIP = \left(\sum_{i=1} \beta_i\right)\left(\sum MEC\right) \tag{5.13}$$

where

β_i is the ith cost factor:

$i = 1$ (for instrumentation)
$i = 2$ (for piping)
$i = 3$ (for installation), etc.

ΣMEC is the cost of major process equipment

This method is described in more detail in Reference [16].

5.2.10 Cost Estimation Method X

This is called the Stallworthy Method [27]. It was developed mainly for petrochemical plants. In some ways the method is similar to Zevnik and Buchanan's approach [28]. The functional units are modified into process steps and the method accounts sidestreams and recycles influence. According to Stallworthy, the battery-limits plant investment can be estimated from the following relationship:

$$PI_{BL} = \frac{Q}{CF}\sum_{1}^{n}[T_f P_f M_f (NS)(SR)] \tag{5.14}$$

where

$$Q = 0.0075$$

PI_{BL} is the battery-limits plant investment. The term battery-limits may be described as a geographical designation for the process area of the project proposed.
T_f is the temperature factor
P_f is the pressure factor
M_f is the material factor
CF is the cost-capacity factor
SR is the stream flow to output stream ratio
NS is the significant process steps in a stream
n is the total number of process streams

The factors temperature, pressure, and material are acquired as number-weighted averages from constituent process steps. The method is accurate within ±15% of a detailed estimate, according to Stallworthy.

5.2.11 Cost Estimation Method XI

This is known as the Hirsch and Glazier Method. The method was proposed by Hirsch and Glazier [29] in 1960. They used the following equation to estimate total battery-limits plant investment:

$$PI_{BL} = (CF_I)\left[EC_Q + IC_A + (TC_{BLQ})\left(\sum_{i=1}^{3} F_i + 1\right)\right] \quad (5.15)$$

where

PI_B is the total battery-limits plant investment
CF_I is the indirect cost factor representing contractor's overhead and profit, supervision, contingencies, and engineering. Normally, this factor is assigned a value of 1.4.
TC_{BLQ} is the total purchased all battery-limits equipment cost calculated on a free-on-board (FOB) basis. This cost does not include any incremental cost for corrosion-resistant alloys.
EC_Q is the all installed equipment cost (e.g., tanks, furnaces, and cooling towers)
IC_A is the incremental cost of corrosion-resistant alloy materials used
F_i is the ith cost factor:

$i = 1$ (for miscellaneous items)
$i = 2$ (for field labor)
$i = 3$ (for piping materials).

The installation cost factors F_1, F_2, and F_3 are defined by the following relationships:

$$F_1 = 0.344 + 0.333 \log TC_{BLQO} + (1.94)(TSC)(TC_{BLQ})^{-1} \quad (5.16)$$

$$\log F_2 = 0.635 - 0.154 \log TC_{BLQO} - 0.992(TC_{BLQ})^{-1}(HXC) + 0.506(TC_{BLQ})^{-1}(FVC) \quad (5.17)$$

$$\log F_3 = -0.266 - 0.014 \log TC_{BLQO} - 0.156(TC_{BLQ})^{-1}(HXC) + 0.556(PDC)(TC_{BLQ})^{-1} \quad (5.18)$$

where

HXC is the heat exchanger cost excluding incremental cost of alloy
FVC is the cost of field-fabricated vessels excluding incremental cost of alloy
$TC_{BLQO} = (TC_{BLQ})/1000$, $k\$$
TSC is the cost of tower shells excluding incremental cost of alloy
PDC is the pump plus driver cost excluding incremental cost of alloy

In Reference [6], this method is demonstrated by solving one numerical example.

5.2.12 Cost Estimation Method XII

This was proposed by Zevnik and Buchanan [28] for fluid-type chemical plants in 1963. The method is based on three factors: process complexity, the number of functional units, and process capacity. All the equipment necessary for performing a process function is called a functional unit. In this method Zevnik and Buchanan acquired the direct investment (per functional unit) from a correlation relating investment to complexity factor, as well as, production capacity. The complexity factor, F_c, is obtained from the following relationship:

$$F_c = 2(10)^{\Sigma_{i=1}^{3} f_i} \tag{5.19}$$

where

f_i is the ith factor:

$i = 1$ (materials)
$i = 2$ (pressure)
$i = 3$ (temperature).

The specified values of f_1 for various construction materials are 0.4 (precious metals), 0 (cast iron, wood, and carbon steel), 0.2 (nickel, monel, inconel, and 300-series stainless steel), 0.3 (hastelloy), and 0.1 (copper, 400-series stainless steel, aluminum, and brass). In order to estimate values of f_2 and f_3, plots are given in Reference [28]. However, the author of Reference [2] present the following relationships for approximating f_2 and f_3, respectively:

$$f_2 = 0.044 \ln Y \tag{5.20}$$

where

$$Y = \begin{cases} \dfrac{P}{14.7}, & \text{for } P > 14.7 \text{ psia} \\ \dfrac{14.7}{P}, & \text{for } P < 14.7 \text{ psia} \end{cases} \tag{5.21}$$

The symbol P represents pressure

$$f_3 = \begin{cases} \dfrac{0.017t}{100} - 0.05, & \text{for } t > 300°\text{K} \\ -\dfrac{0.19t}{100} + 0.6, & \text{for } t < 300°\text{K} \end{cases} \tag{5.22}$$

where t is the temperature

Plots for estimating direct investment, DI, in million dollars per functional unit are presented in Reference [28]. However, this investment can be approximated from the following relationship [2]:

$$DI = \begin{cases} 0.007F_c(AC)^{0.5}, & \text{for } 1 < AC < 10 \\ 0.006F_c(AC)^{0.6}, & \text{for } 10 < AC < 100 \end{cases} \tag{5.23}$$

where

AC is the annual capacity expressed in million pounds per year
F_c is the complexity factor

The total plant investment is expressed by the following equation:

$$TPI = F_{gfu}(n_{fu})(DI)\left(\frac{CE}{102}\right) \tag{5.24}$$

where

TPI is the total plant investment in million dollars
F_{gfu} is the factor for allowance for utilities and general facilities (allocated). The estimated value of F_{gfu} is 1.33.
n_{fu} is the number of functional units
CE is the Chemical Engineering Plant Cost Index for the year in question

5.2.13 Cost Estimation Method XIII

This is known as the Wilson [27] Cost Estimating Technique. In 1971, G.T. Wilson proposed this method for estimating battery-limits investment. The method is derived from the Lang factor as developed by the Miller and Stallworthy methods. The equation developed by Wilson is as follows:

$$PI_{BL} = \left(\prod_{i=1}^{3} f_i\right)(AC_{mp})(IF)MPI) \tag{5.25}$$

where

IF is the investment factor
f_i is the ith factor:

$i = 1$ (materials)
$i = 2$ (pressure)
$i = 3$ (temperature)

AC_{mp} is the main-plant items' mean unit cost
MPI is the total number of main-plant items
PI_{BL} is the battery-limits plant investment

It is to be noted that all major equipment are included in main-plant items with the exception of pumps. Data from 16 plants were considered by Wilson. In most cases estimating accuracies were within $\pm 30\%$.

5.2.14 Cost Estimation Method XIV

This may be called the Probability Cost Estimating Technique, based on a procedure developed for Program Evaluation and Review Technique (*PERT*). This method calls for making three cost estimates: optimistic (lowest cost), most likely (cost), and pessimistic (highest cost) [7]. The following formula is used to estimate mean for a cost element:

$$EC = \frac{LC + 4(ML) + HC}{6} \tag{5.26}$$

where

LC is the lowest cost
HC is the highest cost
ML is the most likely cost
EC is the mean cost of the element

Similarly, the variance, σ^2, of the cost element is given by

$$\sigma^2 = \left(\frac{HC - LC}{6}\right)^2 \tag{5.27}$$

From the Central Limit Theorem (assuming the concerned conditions are satisfied), it follows that the mean of the sum is the sum of the means. Thus, the total expected cost, TEC, is given by

$$TEC = \sum_{i=1}^{k} EC_i \tag{5.28}$$

where

k is the total number of cost elements
EC_i is the expected cost of element i; for $i = 1, 2, 3, \ldots, k$

Similarly, from the Central Limit Theorem we get the variance σ_{TC}^2 of total cost, as follows:

$$\sigma_{TC}^2 = \sum_{i=1}^{k} \sigma_i^2 \tag{5.29}$$

where

k is the total number of cost elements
σ_i^2 is the variance of cost element i; for $i = 1, 2, 3, \ldots, k$

Example 5.6

The cost data given in Table 5.6 are associated with a project having seven cost elements. Calculate the total expected cost and variance of total cost using Equations (5.26)–(5.29).

For cost element No. 1, using the given data in Table 5.6 in Equation (5.26) yields the following mean cost:

$$EC_1 = \frac{10 + 4(12) + 16}{6} = \$12.33k$$

Table 5.6. Cost data.

Cost element no.	Cost element	Optimistic cost (lowest)	Most likely cost	Pessimistic cost (highest)
1	*A*	\$10*k*	\$12*k*	\$16*k*
2	*B*	30	33	44
3	*C*	15	17	24
4	*D*	24	26	28
5	*E*	5	8	15
6	*F*	3	4	5
7	*G*	12	15	25

Table 5.7. Values for mean cost and variance of cost elements presented in Table 5.6.

Cost element no.	Cost element	Mean cost of the element i, EC_i	Variance of the cost element i, σ_i^2
1 ($i = 1$)	A	\$12.33k	1
2	B	34.33	5.44
3	C	17.83	2.25
4	D	26.00	0.44
5	E	8.67	2.78
6	F	4.00	0.11
7	G	16.17	4.69

Similarly, from Equation (5.27) the variance of cost element No. 1 is

$$\sigma_1^2 = \left(\frac{16 - 10}{6}\right)^2$$
$$= 1$$

The other values presented in Table 5.7 for mean cost and variance were obtained in similar manner.

From Table 5.7 and Equations (5.28) and (5.29), we get

$$\begin{aligned} TEC &= EC_1 + EC_2 + EC_3 + EC_4 + EC_5 + EC_6 + EC_7 \\ &= (12.33) + (34.33) + (17.83) + (26.00) + (8.67) \\ &\quad + (4.00) + (16.17) \\ &= \$119.33k \end{aligned}$$

and

$$\begin{aligned} \sigma_{TC}^2 &= (1) + (5.44) + (2.25) + (0.44) + (2.78) + (0.11) + (4.69) \\ &= 16.71 \end{aligned}$$

Thus, the expected total cost of the project is \$119.33k along with variance of total cost equal to 16.71.

5.2.15 Cost Estimation Method XV

This is known as the Cevidalli and Zaidman Method for estimating total cost of utilities in dollars per ton of production. The technique was proposed by Cevidalli and Zaidman in 1980 [30]. The following equation is used to estimate total utility costs:

$$TUC = (0.112)F_{EI}(10)^{-AF}(ICR)^{0.751}(FU)^{0.774} \tag{5.30}$$

$$ICR = \frac{I_{FC}}{AC} \tag{5.31}$$

The symbols used in the above two equations are defined below:

TUC is the total utility cost in dollars per ton of production
F_{EI} is the oil-to-investment-escalation factor
AF is the alloy factor. Its values are presented in section 5.2.12 and in Reference [28]
FU is the total number of functional units in the process
I_{FC} is the fixed capital investment expressed in 10^3 dollars
AC is the yearly production capacity expressed in tons per year

The oil-to-investment-escalation factor, F_{EI}, is defined by

$$F_{EI} = (40.8)\left[\frac{AOC}{ACEI}\right]$$

where

$ACEI$ is the actual Chemical Engineering (CE) Plant Cost Index
AOC is the actual oil cost expressed in \$/bbl

5.3 Sources for Cost Data

This section is concerned with equipment cost data sources. A number of such data sources are listed in Table 5.8 [2].

5.4. Summary

The chapter presented a large number of cost estimation methods and sources for cost data. These methods and sources are selected from the published literature. The cost estimation methods presented in the chapter are as follows:

i) Cost Indexes
ii) Cost-Capacity Factors
iii) Lang Factor

Table 5.8. Selected cost data sources.

Data source no.	Costs for	Cost data source (article/book)
1	pressure vessels	A. Mulet, A.B. Corripio, L.B. Evans, Estimate Costs of Pressure Vessels Via Correlations, Chemical Engineering, Vol. 88, No. 20, 1981, p. 456.
2	heat exchangers	1. G.P. Purohit, Cost of Double Pipe and Multitube Heat Exchangers, Chemical Engineering, Vol. 92, No. 5, 1985, p. 96. 2. D.R. Woods, S.J. Anderson, S.L. Norman, Evaluation of Capital Cost Data: Heat Exchangers, Canadian Journal of Chemical Engineering, Vol. 54, 1976, p. 469. 3. J.D. Kumana, Cost Update on Specialty Heat Exchangers, Chemical Engineering, Vol. 91, No. 13, 1984, p. 164.
3	varieties of process equipment	1. K.K. Humphreys, S. Katell, Basic Cost Engineering, Marcel Dekker, Inc., New York, 1981. 2. R.S. Hall, J. Mately, K.J. McNaughton, Current Costs of Process Equipment, Chemical Engineering, Vol. 89, No. 7, 1982, p. 80. 3. M.S. Peters, K.D. Timmerhaus, Plant Design and Economics for Chemical Engineers, McGraw-Hill, New York, 1980. 4. I.V. Klumpar, S.T. Slavsky, Updated Cost Factors: Process Equipment, Commodity Materials, and Installation Labor, Chemical Engineering, Vol. 92, No. 15, 1985, p. 73.
4	motors, storage tanks, centrifugal pumps, etc.	1. A.B. Corripio, K.S. Chrien, L.B. Evans, Estimate Costs of Heat Exchangers and Storage Tanks Via Correlations, Chemical Engineering, Vol. 89, No. 2, 1982, p. 125. 2. A.B. Corripio, K.S. Chrien, L.B. Evans, Estimate Costs of Centrifugal Pumps and Electric Motors, Chemical Engineering, Vol. 89, No. 4, 1982, p. 115.

Table 5.8. *Continued.*

Data source no.	Costs for	Cost data source (article/book)
5	fabric filters and precipitators	1. S.D. Severson, et al., The Economics of Fabric Filters and Precipitators, Chemical Engineering Progress, Vol. 76, No. 1, 1980, p. 68.
6	corrosion-resistant piping	1. A.H. Tuthill, Installed Cost of Corrosion-Resistant Piping, Chemical Engineering, Vol. 93, No. 4, 6, 8, 1986, pp. 113, 125, 47.
7	air pollution abatement equipment	1. W.M. Vatavak, R.B. Neveril, Estimating Costs of Air Pollution Control Systems, Chemical Engineering, Vol. 87, No. 20, 22, 24, 26, 1980, pp. 165, 157, 111, 7. Also see volumes 88–91.
9	solid waste shredders	1. C.S. Fang, The Cost of Shredding Municipal Solid Waste, Chemical Engineering, Vol. 87, No. 7, 1980, p. 151.
9	distillation columns	1. J.S. Miller, W.A. Kapella, Installed Cost of a Distillation Column, Chemical Engineering, Volume 84, No. 8, 1977, p. 129.
10	—	1. N.P. Chopey, Editor, Modern Cost Engineering: Methods and Data, McGraw-Hill, New York, 1979. This book presents several cost data sources appeared in Chemical Engineering (Journal) from 1970–1980.

iv) Rudd and Watson Method
v) Wessell Method
vi) Hand Factor
vii) Turnover Ratio Method
viii) Factor Method
ix) Ratio Factor Method
x) Stallworthy Method
xi) Hirsch and Glazier Method
xii) Zevnik and Buchanan Method

xiii) Wilson Method
xiv) Probability Cost Estimating Method
xv) Cevidalli and Zaidman Method

The cost indexes include the Chemical Engineering Plant Cost Index, the Marshall and Swift Equipment Cost Index, the Engineering News-Record (*ENR*) Building and Construction Cost Indexes, the Nelson Refinery Construction Cost Index, and the Bureau of Labor Statistics Indexes. All of those indexes are discussed in detail. Periodicals for periodic update of above indexes are listed. A number of sources for cost data are given. These sources are mostly articles published in Chemical Engineering, Canadian Journal of Chemical Engineering, and Chemical Engineering Progress.

5.5 Problems

1. Discuss the following cost indexes:

   i) Marshall and Swift Equipment Cost Index
   ii) Engineering News-Record (*ENR*) Building Cost Index
   iii) Nelson Refinery Construction Cost Index

2. Compare the Chemical Engineering Plant Cost Index with Marshall and Swift Equipment Cost Index.
3. List the values of the Marshall and Swift Equipment Cost Index for the period from 1960 is 1987. Comment on those values.
4. The cost of an equipment manufactured in 1979 was $100,000. For 1979 the Marshall and Swift Index was 599. Calculate the cost of the same equipment in 1984. For 1984 the Marshall and Swift Index was 780. Comment on the end result.
5. What is the range of values for the Cost-Capacity Factor?
6. What is the commonly used value for the Cost-Capacity Factor?
7. Assume that a 2,500 MW power generation plant costs $90 million. Calculate the cost of a similar 4,000 MW power plant, if the Cost-Capacity Factor is 0.8.
8. Compare the following two methods:

   i) The Hand Factor Method
   ii) The Lang Factor method

9. What are the advantages and disadvantages of the Lang Factor Method?

References

1. P.F. Ostwald, P.J. Toole, IE's and Cost Estimating, Industrial Engineering, February 1978, pp. 40–43.
2. T.J. Ward, Cost-Estimating Methods, Modular Instruction Series G: Design of Equipment (Plant design and cost estimating), edited by J. Beckman, Vol. 1, American Institute of Chemical Engineers, New York, 1986, pp. 12–21.
3. J.T. Gallagher, Rapid Estimation of Plant Costs, in Modern Cost-Engineering Techniques, edited by H. Popper, McGraw-Hill Book Company, New York, 1970, pp. 3–10.
4. P.F. Ostwald, Cost Estimating, in Handbook of Industrial Enginerring, edited by G. Salvendy, John Wiley & Sons, New York, 1982, pp. 9.2.1.–9.2.28.
5. J.W. Hackney, Estimating Methods for Process Industry Capital Costs, in Modern Cost-Engineering Techniques, edited by H. Popper, McGraw-Hill Book Company, New York, 1970, pp. 43–58.
6. H.C. Nelson, O.P. Kharbanda, W.A. Janda, J.H. Black, Capital Investment Cost Estimation, in Cost and Optimization Engineering, edited by F.C. Jelen and J.H. Black, McGraw-Hill Book Company, New York, 1983, pp. 321–381.
7. P.F. Ostwald, Cost Estimating for Engineering and Management, Prentice-Hall, Inc., Englewood Cliffs, New Jersey, 1974.
8. M.B. Desai, Preliminary Cost Estimating of Process Plants, Chemical Engineering, July 1981, pp. 65–70.
9. T.H. Arnold, C.H. Chilton, New Index Shows Plant Cost Trends, Chemical Engineering, Vol. 70, Feb. 18, 1963, pp. 143–152.
10. R. W. Stevens, Equipment Cost Indexes for Process Industries, Chemical Engineering, Vol. 54, November 1957, pp. 124–126.
11. K. B. Edmunds, Engineering News-Record Indexes of Cost Trends 1913–68, Engineering News-Record, Vol. 180, March 12, 1968, pp. 77–80.
12. W.L. Nelson, Refinery Construction Cost Indexes, Oil and Gas Journal, Vol. 54, Oct. 1, 1956, pp. 54–60.
13. Monthly Labor Review, Bureau of Labor Statistics, U.S. Department of Labor, Washington, D.C.
14. Consumer Price Index, Bureau of Labor Statistics, U.S. Department of Labor, Washington, D.C.
15. Producer Prices and Price Indexes, Bureau of Labor Statistics, U.S. Department of Labor, Washington, D.C.
16. K.K. Humphreys, Editor, Project and Cost Engineers' Handbook, Marcel Dekker, Inc., New York, 1984, pp. 51–74.
17. E. Seaton, Statistics Canada Bulletin, Vol. 2, June 1973, p. 59.
18. R. W. Stevens, Equipment Cost Indexes, The Cost Engineer, March 1974, pp. 152–157.

19. H. Kolbel, J. Schulze, The Price Index of Chemical Plants, Chemical Industries, Vol. 27, May 1975, pp. 10–15.
20. G. E. Dieter, Engineering Design, McGraw-Hill Book Company, New York, 1983, pp. 324–366.
21. R. C. Chawla, M. M. Varma, Data Sources, Reliability, Updating-Inflation and Relevant Indices, Modular Instruction Series G: Design of Equipment (Plant Design and Cost Estimating), edited by J. Beckman, Vol. 1, American Institute of Chemical Engineers, New York, 1986, pp. 72–84.
22. K.B. Edmunds, *ENR* Indexes Reflect Material and Labor Trends, Engineering News-Record, Vol. 202, March 22, 1979, p. 74.
23. H.J. Lang, Simplified Approach to Preliminary Cost Estimates, Chemical Engineering, Vol. 54, September 1947, pp. 130–133.
24. D.F. Rudd, C.C. Watson, Strategy of Process Engineering, John Wiley & Sons, New York, 1968.
25. W.E. Hand, From Flowsheet to Cost Estimate, Petroleum Refiner, Vol. 37, 1958, pp. 331–334.
26. W.F. Wroth, Factors in Cost Estimation, Chemical Engineering, Vol. 67, October 1960, pp. 204–206.
27. D.H. Allen, R.C. Page, Revised Technique for Predesign Cost Estimating, Chemical Engineering, March 3, 1975, pp. 142–150.
28. F.C. Zevnik, R.L. Buchanan, Generalized Correlation of Process Investment, Chemical Engineering Progress, Vol. 59, February 1963, pp. 70–77.
29. J.H. Hirsch, E.M. Glazier, Estimating Plant Investment Costs, Chemical Engineering Progress, Vol. 56, December 1960, pp. 37–43.
30. G. Cevidalli, B. Zaidman, Evaluate Research Projects Rapidly, Chemical Engineering, Vol. 87, July 14, 1980, pp. 145–152.

CHAPTER 6

Manufacturing, Quality, and Maintenance Costs

6.1 Introduction

Manufacturing, quality, and maintenance costs play an important role in total cost of a product. Therefore, a careful consideration must be given to such costs. The manufacturing cost may simply be described as the sum of fixed and variable costs chargeable to the production of a specified item. Normally this cost (i.e. manufacturing cost) does not include the costs associated with research and development, transportation and distribution, selling, and corporate administration.

In a typical manufacturing company the rework, prevention, waste, appraisal, warranty, and lack of attention costs may account for somewhere between 15 percent and 40 percent of turnover [1]. More specifically, a manufacturer of electronics equipment and systems for Communications and Ocean Engineering use reported an estimate of quality costs as almost 14% of the cost of sales [2]. Nevertheless, one may say that the quality costs provide an economic means through which company management and those practicing quality control can communicate in business terms effectively. These costs (quality costs) cross department lines by involving company activities such as design, purchasing, service, and manufacturing.

The maintenance cost may be stated in broader terms as the labor and materials expense needed to maintain items in suitable use condition. The life cycle cost studies associated with engineering systems have revealed the significant economic impact of maintenance activities. These activities, in military systems, can be responsible for up to as high as 70% of life cycle costs.

A significant amount of published literature is available on each of those three costs: manufacturing, quality, and maintenance. This chapter presents such costs separately.

6.2 Manufacturing Costs

The manufacturing costs account for a significant proportion of life cycle costs of engineering products. The cost of goods manufactured can be broken down into following elements:

i) Direct material cost
ii) Direct labor cost
iii) Indirect material cost
iv) Indirect labor cost
v) Manufacturing expense

In regard to estimating manufacturing (production) cost one can say without any hesitation that it is the most developed procedure relative to others. The job description or plan is required to estimate manufacturing costs. The manufacturing plan should provide information on items such as those given below:

i) Production quantity required
ii) Manufacturing personnel required
iii) Schedules
iv) Assembly sequence diagram
v) Major tooling
vi) Test equipment

The plan should be prepared by the manufacturing department of the company.

6.2.1 Estimating Manufacturing Costs for New Processes

In estimating manufacturing costs for new processes one has to look into many cost items. This section presents a list of such items which should not be overlooked when estimating manufacturing costs for new processes. Otherwise, an incorrect estimate of manufacturing costs may result. The cost items are as follows [3]:

i) Maintenance and repair
ii) Labor
iii) Utilities
iv) Packaging and shipping
v) Raw materials
vi) Royalties (if applicable)
vii) Direct overhead (usually plant supervision)
viii) Laboratory charges (process control and quality control work)

ix) Factory supplies (housekeeping supplies, wiping clothes, instrument charts, etc.)
x) Development costs (if applicable)
xi) Plant indirect overhead (e.g., plant office expense)
xii) Property taxes
xiii) Depreciation
xiv) Insurance

Items (i)–(x) are related to direct cost and the remaining four to indirect cost. Each of those items is described in detail in Reference [3].

6.2.2 Methods for Estimating Manufacturing Costs

This section presents two quick approximation methods for estimating manufacturing costs. These are the guidelines for estimating operating costs and the 1-3-9 rule method. Both these techniques are presented below:

Method I: Guidelines for Estimating Operating Costs in the Absence of Company Records

These guidelines can be used for estimating operating (manufacturing) cost when accessibility to company records is not possible. In addition, the guidelines are quite useful for developing initial estimates in the early phases of a project. The three major components of the operating cost are

i) Direct production cost
ii) Indirect cost, and
iii) Distribution cost.

The direct production cost is composed of cost elements such as labor, materials, utilities, waste disposal, maintenance, royalties, supervision, operating supplies, contingencies, laboratory, and payroll charges. The cost estimates for labor, materials, and utilities can be obtained from manning tables, price lists, and literature or similar operations, respectively. Similarly, the cost of waste disposal can be estimated from literature or similar operations. The remaining seven cost elements [4] can be estimated using the following guidelines, respectively:

i) 2–10 percent of annual investment
ii) 1–5 percent of sales

iii) 10–25 percent of labor
iv) 0.5–1.0 percent of annual investment
v) 1–5 percent of direct costs
vi) 10–20 percent of annual labor cost
vii) 30–45 percent of labor plus supervision cost

More specifically, the above seven guidelines are associated with maintenance, royalties, supervision, operating supplies, contingencies, laboratory, and payroll charges, respectively. The elements of the indirect cost are the insurance, depreciation, interest, general plant overhead, and real estate taxes. These items can be estimated using the guidelines given below, respectively:

i) 0.5–1.0 percent of annual investment
ii) 5–10 percent of annual investment
iii) 10–12 percent of annual investment
iv) 50–70 percent of maintenance, labor, and supervision costs
v) 1–2 percent of annual investment

Finally, the components of the distribution cost are packaging and shipping. The cost of packaging can be determined from container costs and the cost of shipping can be taken as 1–3 percent of sales.

Method II: The 1-3-9 Rule

This rule can be used to acquire fast approximations of manufacturing costs [5]. The material cost level of the product, being manufactured, serves as base for the estimate of manufacturing cost. The cost of the material, *CM*, is expressed as follows:

$$CM = (\alpha)(\beta)(CW)(CRM) \tag{6.1}$$

where

CW is the component's weight
CRW is the cost of raw material/unit of weight

The symbols α and β represent tooling factor and waste factor, respectively.

In this case the physical volume of the part is estimated by the user using the engineering drawing. The weight of the part is then acquired by multiplying the material density [6]. One should note here that in obtaining estimates for the material cost, the values of both the tooling and waste factors are augmented above unity by the magnitude of their respective allowances. For example, the tooling and waste allowances of 8% and 9%, respectively, the values of tooling and waste factors become 1.08 (i.e., $\alpha = 1.08$) and 1.09 ($\beta = 1.09$), respectively. Thus from equation (6.1) we get

$$\begin{aligned} CM &= (1.08)(1.09)(CW)(CRM) \\ &= (1.1772)(CW)(CRM) \end{aligned} \tag{6.2}$$

The manufacturing cost of a part fabricated internally can be estimated from the following relationship:

$$MFC_I = 3(CM) \tag{6.3}$$

where MFC_I is the manufacturing cost of a part fabricated internally. Similarly, the manufacturing cost of a part purchased from an outside vendor is expressed as

$$MFC_V = 4(CM) \tag{6.4}$$

where MFC_V is the manufacturing cost of a part procured from an outside vendor. The selling price, SP_I, of the unit (internally fabricated part) can be established from the following relationship:

$$\begin{aligned} SP_I &= (3)MFC_I \\ &= 3(3CM) \\ &= 9(CM) \end{aligned} \tag{6.5}$$

6.2.3 The Manufacturing Progress Curve

This curve is also known as a learning curve. Since the early contributions of T.P. Wright [7] in 1936, there have been a significant number of empirical evidences gathered from manufacturing situations concerning the learning process [8, 9]. These evidences state that as the workers involved attain experience in their jobs, their effort required to accomplish a unit of production will diminish by a constant percentage each time the production becomes twofold. The extent and rate of time reduction are governed by several factors. Some of those are as follows:

i) Worker/management relationships
ii) Lengths of the production runs
iii) The nature of the production process
iv) The degree of preproduction planning
v) Product design standardization

The manufacturing progress curve (learning curve) is defined by the following equation:

$$PE = E_f Z^{\beta} \tag{6.6}$$

where

PE is the production effort given in either cumulative average hours for production of Z units or hours per unit
E_f is the effort (hours) needed to produce the first unit
Z is the cumulative total of units produced
β is the (negative) slope parameter of the learning curve. This parameter determines the percentage value by which PE diminishes each time the value of Z becomes twofold.

Normally, the learning curve is constructed for a doubling curve of production. However, there is no restriction to construct it for a tripling situation or for any amount. The case of doubling of production is described earlier. However, for a typical learning curve percentage of 80% we describe it again. Each time the cumulative production quantity is doubled, the time required to manufacture is 80% of what it was prior to the occurrence of doubling. For example, consider production quantities Z_1 and Z_2 (for doubling of production $Z_2 = 2Z_1$) and their corresponding production times PE_1 and PE_2. For the improvement rate of 80%, the PE_2 is given by $PE_2 = 0.8PE_1$. The same argument applies for other improvement rates, say, 85%, 75%, and 90%. For given values of improvement rates, the value of β can be estimated as follows:

Assume that production times for the first and second units, respectively, are

$$PE_1 = E_f Z_1^{\beta} \tag{6.7}$$

and

$$PE_2 = E_f Z_2^{\beta} \tag{6.8}$$

Dividing Equation (6.7) by Equation (6.8) results in

$$\frac{PE_1}{PE_2} = \frac{Z_1^{\beta}}{Z_2^{\beta}}$$

$$= \left(\frac{Z_1}{Z_2}\right)^{\beta} \tag{6.9}$$

For $Z_2 = 2Z_1$ (i.e., because of doubling of production), the above equation becomes

$$\frac{PE_1}{PE_2} = \left(\frac{Z_1}{2Z_1}\right)^{\beta} \tag{6.10}$$

Therefore,

$$\frac{PE_2}{PE_1} = 2^{\beta} \tag{6.11}$$

For an improvement rate of 70% (i.e., $PE_2 = 0.7PE_1$) the above equation yields

$$\frac{0.7PE_1}{PE_1} = 2^{\beta} \tag{6.12}$$

Therefore,

$$\beta = \frac{\ln(0.7)}{\ln(2)} = -\frac{0.3567}{0.6931}$$
$$= -0.5146$$

Thus, when the improvement rate is 70%, the value of β is -0.5146. Similarly, for an improvement rate of 80% (i.e., $PE_2 = 0.8PE_1$), the Equation (6.11) yields

$$\frac{0.8PE_1}{PE_1} = 2^{\beta} \tag{6.13}$$

Therefore,

$$\beta = \frac{\ln(0.8)}{\ln(2)} = -\frac{0.2231}{0.6931}$$
$$= -0.3219$$

Thus, the β takes the value of -0.3219 when the improvement rate is 80%.

Similarly, one can obtain values of β for other specified improvement rates as shown in Table 6.1.

Example 6.1

Assume that the first of a lot of 30 devices takes 250 hours to manufacture. If the improvement rate for doubling of production is 80%, calculate the time to produce a twentieth device.

Table 6.1. Estimated values for β for selected learning curve percentages.

Learning curve percentage	The estimated value for β
90	−0.152
85	−0.234
82	−0.286
80	−0.322
79	−0.340
75	−0.415
70	−0.515
67	−0.578
65	−0.622
60	−0.737

From Table 6.1 for the specified learning curve percentage (80%), the value of β is -0.322. The given values of E_f and Z are 250 hours and 20, respectively.

Using the above given data into Equation (6.6) yields

$$PE = 250(20)^{-0.322}$$

$$= 95.28 \text{ hours}$$

Thus, the twentieth device will take 95.28 hours to complete.

6.2.4 Use of Least-Squares in Learning Curve Work

A situation may occur in which one has to specify the learning curve for a given set of data. In such a situation, the method of least squares can be utilized. The method in question specify the line of best fit for the given set of data. Equation (6.6) can be expressed as a logarithmic straight line, as follows [10, 11]:

$$\log PE = \log E_f + \beta \log Z \tag{6.14}$$

The least-squares method leads to

$$\log E_f = \frac{\Sigma A \, \Sigma B^2 - \Sigma B \, \Sigma A B}{K \, \Sigma B^2 - (\Sigma B)^2} \tag{6.15}$$

where

$A = \log PE$
$B = \log Z$, and the symbol K denotes the number of data points.

$$\beta = \frac{K \Sigma B A - \Sigma B \Sigma A}{K \Sigma B^2 - (\Sigma B)^2} \tag{6.16}$$

This method is demonstrated through the following example:

Example 6.2

Assume that a set of data contains six data points as shown in Table 6.2. Determine the learning function that best fits the given data. In addition, estimate the value of the learning curve percentage. Using the data given as Table 6.2, we developed Table 6.3. In this example the value of K is equal to 6. Using the end results of Table 6.3 in Equation (6.15), we get

$$\log E_f = \frac{(12.3542)(8.2464) - (6.3216)(12.5082)}{6(8.2464) - (6.3216)^2}$$

$$= \frac{101.8777 - 79.0718}{9.5158}$$

$$= 2.3966$$

Therefore, $E_f = 249.25$ man-hours. Similarly, using the end results ofTable 6.3 in Equation (6.16) leads to

$$\beta = \frac{6(12.5082) - (6.3216)(12.3542)}{6(8.2464) - (6.3216)^2}$$

$$= -\frac{3.0491}{9.5158}$$

$$= -0.3204$$

The learning curve percentage, θ, is given by

$$\theta = 2^{\beta}(100) \tag{6.17}$$

By substituting the calculated value of β, the above equation yields

$$\theta = 2^{-0.3204}(100)$$

$$= 80.08\%$$

Thus, the learning curve percentage is 80.08.

Table 6.2. Specified data points.

Unit number (Z)	Unit labor man-hours (PE)
2	200
4	160
8	128
16	102
32	82
64	66

Table 6.3. Calculations for least-squares fit.

Unit number (Z)	Unit labor man-hours (PE)	A (log PE)	B (log Z)	$A \cdot B$	B^2
2	200	2.3010	0.3010	0.6926	0.0906
4	160	2.2041	0.6021	1.3271	0.3625
8	128	2.1072	0.9031	1.9030	0.8156
16	102	2.0086	1.2041	2.4186	1.4499
32	82	1.9138	1.5051	2.8805	2.2655
64	66	1.8195	1.8062	3.2864	3.2623
		$\Sigma A = 12.3542$	$\Sigma B = 6.3216$	$\Sigma A \cdot B = 12.5082$	$\Sigma B^2 = 8.2464$

6.2.5 Estimating Item Cost Involving Learning

In this case we are concerned with determining average cost per unit involving learning [12]. In order to develop an equation for that cost, consider the manufacture of a group of M units. For the manufacture of these units, the direct labor-hours needed are given by

$$H_M = \sum_{z=1}^{M} PE_z \tag{6.18}$$

where

M is the number of units in the lot
H_M is the cumulative number of direct labor-hours
PE_z is the direct labor-hours needed to manufacture unit z

The cumulative number of direct labor-hours can be approximated from the following relationship:

$$H_M \cong \int_0^M PE_z \, dz \tag{6.19}$$

Substituting for PE_z from Equation (6.6), Equation (6.19) becomes

$$\begin{aligned} H_M &\cong \int_0^M E_f z^\beta \, dz \\ &\cong E_f \left[\frac{z^{\beta+1}}{\beta+1} \right]_0^M \\ &\cong E_f \frac{M^{\beta+1}}{(\beta+1)} \end{aligned} \tag{6.20}$$

In order to obtain the cumulative average number of direct labor-hours, H_{ca}, we divide Equation (6.20) by M, as follows:

$$\begin{aligned} H_{ca} &\cong \frac{H_M}{M} \cong \frac{E_f \cdot M^\beta \cdot M}{(\beta+1)M} \\ &\cong \frac{E_f M^\beta}{(\beta+1)} \end{aligned} \tag{6.21}$$

Thus, the product cost per unit for product j is expressed as below:

$$C_{pj} = C_{DL} + C_{DM} + C_{OH} \tag{6.22}$$

where

$$C_{DL} = H_{ca} \cdot r = \frac{E_f M^\beta \cdot r}{\beta + 1} \tag{6.23}$$

$$C_{OH} = (H_{ca})(r)(r_{OH}) = \frac{E_f M^\beta r \cdot r_{OH}}{\beta + 1} \tag{6.24}$$

The symbols used in above equations are defined as follows:

C_{DL} is the direct labor cost
C_{DM} is the direct material cost per unit
C_{OH} is the overhead cost
r is the hourly rate for direct labor
r_{OH} is the overhead rate given as a decimal fraction of the hourly rate for direct labor

Example 6.3

Assume that in a manufacturing set-up 40 units are to be manufactured and the learning curve percentage is 90% (for doubling production). The first unit will take 400 direct labor-hours to manufacture. The estimated values of C_{DM}, r, and r_{OH} are \$1,000, \$15 per hour, and 0.6, respectively. Calculate the total cost to produce 40 units.

For the given learning curve percentage of 90%, the value of β from Table 6.1 is -0.152. Thus, substituting the above data into Equation (6.22) yields

$$C_{pj} = \frac{(15)(400)(40)^{-0.152}}{(1-0.152)}[1+0.6]+1,000$$

$$= 6,461.93 + 1,000$$

$$= \$7,461.93$$

Thus, the average cost per unit is \$7,461.93. The total cost, C_T, for the lot of 40 units is

$$C_T = (\$7,461.93(40)$$

$$= \$298,477.2$$

For the assumed purchasing cost, C'_{pj}, of a unit, one can easily determine from Equation (6.18) the number of units that must be manufactured for a break-even unit cost. Thus, in Equation (6.22) we replace C_{pj} with C'_{pj} and solve it for M to get [13]

$$M = \left[\frac{\{C'_{pj} - C_{DM}\}(\beta + 1)}{E_f(1 + r_{OH})r}\right]^{1/\beta} \tag{6.25}$$

Example 6.4

Assume that a unit manufactured in Example 6.3 can be bought from an outside company for \$6,000. Using this and other data given in Example 6.3, determine the quantity of units that have to be manufactured so that the manufacturing alternative becomes as good as purchasing from outside.

By substituting the given data into Equation (6.25), we get

$$M = \left[\frac{\{6,000 - 1,000\}(1 - 0.152)}{400(1 + 0.6)(15)}\right]^{-1/0.152}$$

$$= \left[\frac{(5,000)(0.848)}{9,600}\right]^{-1/0.152}$$

$$= 216 \text{ units}$$

Thus, one has to produce 216 units to break-even.

6.2.6 Mathematical Models Used in Estimating Manufacturing Costs

Model I

This model is concerned with estimating direct material cost [10]. The direct material cost, DMC, in dollars, of a unit is defined by

$$DMC = (WT)(MP)\left(1 + \sum_{i=1}^{3} \gamma_i\right) - SP \qquad (6.26)$$

where

SP is the unit price of expected material salvage, expressed in dollars per unit
MP is the material price expressed as per pound, per volume, or per linear foot
WT is the weight of a unit usually expressed in pounds
γ_i is the ith losses expressed in decimals:

$i = 1$ (due to shrinkage)
$i = 2$ (due to scrap)
$i = 3$ (due to waste).

Model II

This model can be used to estimate single point tools' lives. The equation for the model is expressed as follows [14]:

$$S_c = KL^{-m} \qquad (6.27)$$

where

m and K are the equation's slope on a log-log plot and a constant, respectively.
S_c and L are the cutting speed (feet per minute) and the life of the tool (minutes), respectively.

Equation (6.27) can be expressed as a logarithmic straight line of the following form:

$$\log S_c = \log K - m \log L \qquad (6.28)$$

Typical estimates for K and m may be found in various metal cutting handbooks.

Model III

This model is concerned with developing mean unit cost estimate, *CM*, for a single-point rough-turning operation. The equation is presented below [10]:

$$CM = \sum_{i=1}^{4} C_i \tag{6.29}$$

where

C_1 is the handling cost
C_2 is the tool cost
C_3 is the machining cost
C_4 is the tool-changing cost

For estimating C_1, C_2, C_3, and C_4, equations are presented below. The material handling cost, C_1, is given by

$$C_1 = (HT)(COT) \tag{6.30}$$

where

HT is the handling time per work piece expressed in minutes
COT is the operating time cost expressed in \$/minute

The tool cost is defined by

$$C = (TC) \cdot (MT)/MLT \tag{6.31}$$

where

MT is the work piece machining time expressed in minutes/piece
MLT is the mean tool life given in minutes
TC is the tool cost expressed as dollars per cutting edge

The machining cost is expressed as follows:

$$C_3 = (COT)(MT) \tag{6.32}$$

Both COT and MT were defined earlier. Finally, the tool-changing cost is given by

$$C_4 = (COT)(MT)(TCT)/MLT \tag{6.33}$$

where TCT is the tool-changing time expressed as minutes per operation. The rest of the symbols COT, MT, and MLT were defined earlier.

Model IV

This model is concerned with estimating the machining cost. This cost is defined [15–17] by

$$C_M = (CR) \cdot (PT)$$
$$= \frac{1}{60}\left[\frac{(MC)(1+R_m)}{100} + \frac{(LR)(1+R_0)}{100}\right][T_m + IT] \qquad (6.34)$$

where

CR is the cost rate expressed in dollars per minute
PT is the production time for a unit
T_m is the machining time
MC is the machine cost expressed in dollars per hour
LR is the labor rate for operator expressed in dollars per hour
IT is the idle or non-production time
R_0 is the operator overhead rate expressed in percentage
R_m is the machine overhead rate expressed in percentage

Model V

This model is concerned with estimating unit tooling cost with group technology application. For many years in several countries the group technology has been in use in many forms and degrees. It can simply be described as a manufacturing concept that identifies and exploits the likeness of components and operation processes during both design and manufacture [18]. The group technology application helps to reduce cost (in this model the tooling cost). The unit tooling cost, TC_g (\$/piece), through group tooling approach is expressed as follows [18]:

$$TC_g = \frac{\left\{CGJ + \sum_{j=1}^{k} AC_j\right\}}{\theta} \qquad (6.35)$$

where

θ is the number of parts manufactured
CGJ is the group jig or fixture cost expressed in dollars
k is the number of adapters employed for the manufacture of a group of components or parts
AC_j is the adapter j cost expressed in dollars

Model VI

This model is concerned with estimating time required to accomplish a cut on a lathe. The cutting time, CT, in minutes is defined [14, 19, 20] by

$$CT = \frac{CL}{FR} \tag{6.36}$$

where

FR is the feed rate expressed in inches/spindle revolution
CL is the cut length expressed in units of inches

Model VII

This model is concerned with determining the tool cost associated with a cutting tool that is brazed to the toolholder. This cost, CT, is expressed [15] by

$$CT = \frac{C_{rs}\mu + C_{TL}}{B} \tag{6.37}$$

where

$B = (\mu + 1)$
C_{rs} is the resharpening cost expressed in dollars
C_{TL} is the cost of tool expressed in dollars
μ is the total number of resharpenings

However, for a throwaway (insert) tool, the cost, CT, is expressed as below:

$$CT = \sum_{i=1}^{2} D_i \tag{6.38}$$

where

$D_1 = C_{ti}/m_{ce}$
$D_2 = C_{th}/m_{ceh}$
C_{th} is the toolholder's cost
C_{ti} is the cost of tool insert expressed in dollars
m_{ce} is the (total) number of cutting edges
m_{ceh} is the total number of cutting edges in toolholder's entire life

Model VIII

This model is concerned with estimating operational cost for manufacturing direct labor. Thus, the total direct labor cost, *DLC*, on a per operation, subassembly, or unit basis is given by

$$DLC = \sum_{j=1}^{N} PHR_j \left(STD_j + \frac{SSO_j}{LS} \right) \tag{6.39}$$

where

$j = 1, 2, 3, 4, \ldots, N$ operation
LS is the lot size
PHR_j is the operator's hourly rate in dollars
SSO_j is the setup standard for the operation expressed in hours per lot
STD_j is the operation's estimated standard time given in hours per unit

The model is described in detail in Reference [10].

Model IX

This model is concerned with determining the time, *TM*, in minutes needed for machining a specified surface. The time in question is defined [14] by

$$TM = \frac{WD}{(FI)(MS)} \tag{6.40}$$

where

FI is the feed increment (inches/stroke)
WD is the width of the item to be cut expressed in inches
MS is the machine strokes in a minute needed to produce a cutting speed of CS surface feet per minute

Model X

This model is concerned with estimating the material removal rate associated with turning operations on lathes [14]. The knowledge of this rate is necessary for estimating cutting times. The material removal rate is expressed by

$$R_{mr} = (FR)(DP) \cdot S_k \tag{6.41}$$

where

FR is the feed rate expressed in inches/revolution of the spin
DR is the cut depth expressed in inches
S_k is the cutting speed expressed in surface feet/minute

6.3 Quality Costs

These costs represent an important component of sales turnover. According to a brochure [21] prepared by the British Department of Trade and Industry, the quality costs account for between 5 and 25% of turnover. Nevertheless, one can state that these costs provide an economic means through which company management and those practicing quality control can communicate in business terms effectively. The history of the quality cost concept can be traced back to the early fifties when works entitled "The Economics of Quality" and "The Economic Choice of Sampling Systems in Acceptance Sampling" were published by Juran [22] and Sittig [23], respectively. The works of Feigenbaum [24] and Freeman [25], published in the early sixties, are regarded as among the earliest publications which classified quality costs into four categories: appraisal cost, internal failure cost, prevention cost, and external failure cost. Another major milestone in the history of quality costs took place in 1961, when the American Society for Quality Control (ASQC) established the "Quality Cost Technical Committee". Two years later in 1963, the United States Department of Defense issued MIL-Q-9858A entitled "Quality Program Requirements". This document is considered to have played an important role in the development of the quality cost concept as it made "Costs Related to Quality" a requirement on concerned military contractors and subcontractors.

Ever since then several other developments have taken place [26] and a large number of publications relating to quality costs have appeared [27].

6.3.1 Categories of Quality Costs

There are four categories of quality costs: internal failure cost, prevention cost, external failure cost, and appraisal cost [28]. Each of these costs are discussed below, separately.

Internal Failure Cost

This cost occurs when items fail to satisfy quality requirements before transfer of ownership (of items involved) to the customer. Some of the subcategories of the internal failure cost are shown in Figure 6.1. The repair cost incurred to repair defective items so that they fully satisfy specified requirements. The scrap cost results from the defective item which has to be abandoned because it can not be repaired or used economically. Failure analysis cost incurs when finding out the cause of item failure. Finally, the reinspection and retest costs are associated with reinspection or retest of items that have been repaired.

Prevention Cost

This cost is associated with planning, implementing, and maintaining the quality system. The purpose of the quality system is to assure conformance to quality specifications at economic levels. The prevention cost is expressed by

$$PC = QEC + QDEC + QPC + QTC \tag{6.42}$$

where

PC is the prevention cost

QEC is the cost of quality engineering. This cost is associated with the development and implementation of the overall quality plan, the inspection plan, and so on.

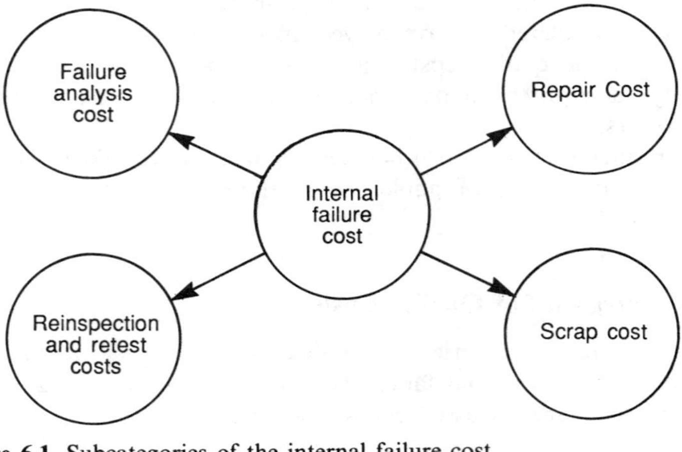

Figure 6.1. Subcategories of the internal failure cost.

QDEC is the cost associated with design and development of quality measurement and control equipment
QPC is the cost of quality planning by functions other than quality control
QTC is the cost associated with quality training. This cost includes the cost of developing and maintaining quality training programs.

External Failure Cost

This cost occurs when items do not satisfy quality specifications after transfer of their ownership to customers. The total external failure cost is given by

$$EFC = RC + CC + WC + REC \tag{6.43}$$

where

RC is the cost associated with the repair of returned items
CC is the cost associated with the adjustment of complaints
WC is the cost of warranties
REC is the cost of handling and replacement of rejected (returned) items

Appraisal Cost

This cost is associated with determining the degree of conformance to quality specifications. This cost has several elements and some of those are as follows [29]:

i) Cost of testing and inspection of incoming material
ii) Cost of testing and inspection of items being produced
iii) Cost of calibration and maintenance of equipment used for the purpose of evaluating quality
iv) Cost of conducting product quality audits
v) Cost of materials and services consumed in testing

6.3.2 Distribution of Quality Costs

In this section we report the distribution of quality costs in the banking and the electronic equipment manufacturing industries. According to Reference [30], the quality costs account for 25% of a bank's total operating costs. In addition, the estimates of distribution among appraisal, internal failure (error), prevention, and external failure (error) costs are given in Figure 6.2. According to Reference

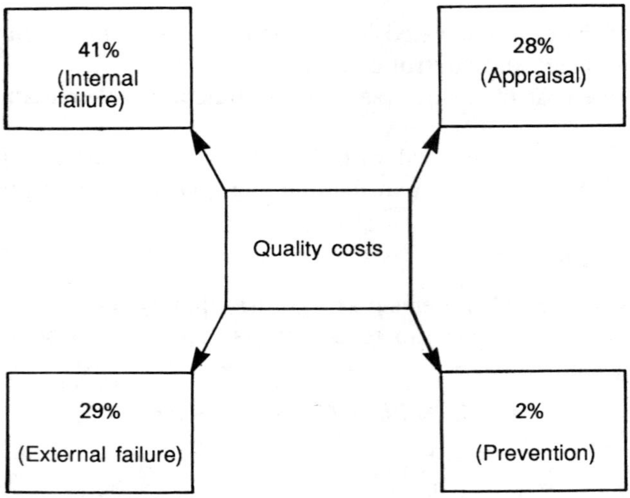

Figure 6.2. Distribution of quality costs in the banking industry.

[2] the approximate estimate of quality cost in Hermes Electronics Limited were in order of 14% of the sales. The total quality cost was distributed as follows:

i) Internal failure cost: 45.5% (of total quality cost)
ii) Appraisal cost: 36.1%
iii) External failure cost: 12.7%
iv) Prevention cost: 5.7%

Further breakdowns for appraisal, internal failure, prevention, and external failure costs were as below.

Appraisal Cost:

i) In-process inspection and test: 19.5%
ii) Equipment calibration: 7.7%
iii) Receiving inspection: 4.6%
iv) Others: 4.3%

Internal Failure Cost:

i) Rework: 11.5%
ii) Scrap: 11.5%
iii) Troubleshooting and failure analysis: 6.8%
iv) Remedial engineering: 6.1%

v) Reinspection and retest: 5.7%
vi) Material review: 2.9%
vii) Others: 1.0%

Prevention Cost:

i) Quality planning: 2.2%
ii) Others: 2.1%
iii) Quality data: 1.5%

External Failure Cost:

i) Rework, reinspection, and retest: 6.4%
ii) Sample replacement: 2.7%
iii) Failure analysis: 2.3%
iv) Others: 1.3%

6.3.3 Reducing Quality Costs

In order to improve quality and reduce quality cost, the strategic planning is one approach which can be used. The approach is composed of following six steps [31]:

i) Review current conditions and past performance
ii) Appraise the environment
iii) Establish objectives
iv) Formulate and choose a strategy
v) Implement the program
vi) Report and review the plan

The above six steps are described in detail in Reference [31].

6.3.4 Quality Cost Related Indexes

This section presents four selective indices used in quality cost work.

Index I

This index can be used to measure quality cost effectiveness of the entire company. The index is defined below [28]:

$$QCE = \frac{TFC}{TAC + TPC} \tag{6.44}$$

where

QCE is the quality cost effectiveness index
TFC is the total failure cost
TAC is the total appraisal cost
TPC is the total prevention cost

Index II

This index is concerned with evaluating vendor quality cost performance. The quality cost performance index, $QCPI$, is defined [33] by

$$QCPI = \frac{VQC + PC}{PC} \tag{6.45}$$

where

VQC is the vendor quality cost
PC is the purchased cost

For a perfect supplier, the value of $QCPI$ will be unity. However, the values of the index from 1.0 to 1.009, 1.010 to 1.039, 1.040 to 1.069, and 1.070 to 0.099 may be intepreted as excellent, good, fair, and poor, respectively. The value of $QCPI$ equal to 1.1 or above indicates that immediate corrective action is required.

Index III

This is defined by

$$QCI = \frac{QC}{VO}(100) + 100 \tag{6.46}$$

where

QCI is the quality cost index
QC is the quality cost
VO is the value of output

When quality costs are negligible, the value of this index is 100. However, in reality the value of QCI equal to 105 can readily be attained. For companies which ignore quality costs, the value of QCI often varies between 110 to 130.

Index IV

This index can be used to examine each step of the manufacturing test process. The index, I, is defined below [30]:

$$I = \frac{FC}{IHC}$$

$$= \frac{MCR_1 + (MFT) \cdot (MLR_1)}{MCR_2 + [MLR_2/MDD]}$$

where

FC is the field cost per defect
IHC is the in-house cost per defect
MCR_i is the ith mean cost of replacement part:

$i = 1$ (in field)
$i = 2$ (in factory).

MLR_i is the ith mean labor and burden rate:

$i = 1$ (in field)
$i = 2$ (in factory).

MFT is the mean repair time in the field
MDD is the mean number of defects discovered and rectified per test hour at the operation being reviewed

If the value of I for any operation is less than unity, the test operation is to be stopped as not being economical. This index is described in detail in Reference [30].

6.4 Maintenance Costs

One of the most perplexing problems in life cycle cost estimating is to accurately estimate maintenance costs. There are several maintenance cost factors. Most of the typical ones that impact cost of ownership are given below [33]:

i) Cost of facilities
ii) Cost of requirements for special tools
iii) Cost of repair parts
iv) Cost of labor used in maintenance
v) Cost of labor personnel training
vi) Cost of equipment inventories
vii) Cost of equipment publications used for maintenance and training

Each of the above factors is described in detail in Reference [33].

There are several models and approaches available in the published literature for estimating various types of maintenance costs. This section presents some of them.

6.4.1 Estimating and Controlling Plant Maintenance Costs

For estimating and controlling plant maintenance costs, there are four general methods available. These are i) developing repair cost estimates for each piece of equipment; ii) making use of a percentage of investment costs; iii) relating maintenance to power requirements; and iv) making use of company records on the same or similar operations [34].

For estimating annual maintenance expense for several different plants the following equations are presented from References [34–36]:

$$AMC = 0.011A - 300{,}000 \tag{6.48}$$

where

$$A = \sum_{i=0}^{m} ST_i y_i$$

$$AMC = 0.004A - 83{,}091 \tag{6.49}$$

$$AMC = 0.009A + 149{,}000 \tag{6.50}$$

$$AMC = 0.083B \tag{6.51}$$

where

$$B = \sum_{i=0}^{m} (ST_i y_i)/EL_i$$

$$AMC = 0.068D \tag{6.52}$$

The definitions of the symbols used in above equations are given below:

AMC is the annual maintenance cost
m is the number of units being maintained
ST_i is the investment in a plant unit i
y_i is the unit i time in years from its installation
EL_i is the unit i total life

The above five Equations (6.48), (6.49), 6.50), (6.51), and (6.52) are for cement plants, coke plants, pulp and paper plants, electrolytic plants, and silicone products plants, respectively. Before using any of these equations in real life applications, one should consult References [34–36] for more detail.

6.4.2 Estimating Corrective Maintenance Cost

Estimating corrective maintenance cost is an important task. For this purpose, the following relationship can be used:

$$CMC = (T_{so}) \cdot (MLC)\left[\frac{MTTR}{MTBF}\right] \tag{6.53}$$

where

CMC is the annual cost of corrective maintenance
$MTTR$ is the mean time to repair
$MTBF$ is the mean time between failures
T_{so} is the scheduled operating hours
MLC is the maintenance labor cost per hour

Example 6.5

Assume that the annual scheduled operating hours for a system are 2,400 hours. The estimated mean time between failures ($MTBF$) and mean time to repair ($MTTR$) of the system are 600 hours and 30 hours, respectively. The cost of maintenance labor is $10 per hour. Calculate the annual cost for corrective maintenance. Substituting the above data into Equation (6.53) results in

$$\begin{aligned} CMC &= (2{,}400)(10)\left[\frac{30}{600}\right] \\ &= \$1{,}200 \end{aligned}$$

Thus, the estimated annual cost for system corrective maintenance is $1,200.

6.4.3 Estimating Logistic Support Cost

One model for this purpose was developed by the United States Air Force. In this model, the system support cost is expressed as sum of support costs for the subsystems plus costs of spare engines, new facilities, and fuel consumption. More specifically, it is expressed as follows [37]:

$$SSC = k_f + k_{se} + k_{nf} + k_{tp} + k_{mtd} + k_{irs} + k_{oem} + k_{fem} + k_{is} + k_{se} \tag{6.54}$$

where

SSC is the system support cost
k_f is the fuel consumption cost
k_{se} is the cost of spare engines
k_{nf} is the new facilities cost
k_{tp} is the training equipment plus personnel training cost
k_{mtd} is the management plus technical data cost
k_{irs} is the initial or replacement spares cost
k_{oem} is the cost of on-equipment maintenance
k_{fem} is the cost of off-equipment maintenance
k_{is} is the cost of inventory entry plus supply management
k_{se} is the support equipment cost

6.4.4 Estimating Fire Control Radar's Maintenance Cost

In Reference [38], the maintenance cost is expressed as follows:

$$FCRMC = (C_{fm} T_y Y)/1{,}000 \tag{6.55}$$

where

$FCRMC$ is the fire control radar's maintenance cost
T_y is the flying hours per year
Y is the number of years in operation

The natural logarithm of maintenance cost, C_{fm}, per flying hour per unit in 1974 dollars ($\times 10^3$) is given by the following relationship:

$$\ln C_{fm} = A_1 + B_2 \ln P \tag{6.56}$$

where

$A_1 = -2.086$
$B_2 = 0.611$
$\ln C_{fm}$ is the natural logarithm of C_{fm}.
P is the peak power expressed in kilowatts.

6.4.5 Estimating Doppler Radar's Maintenance Cost

This cost can be estimated from the following expression:

$$DRMC = [(C_{adm}) \cdot (Y)]/1{,}000 \tag{6.57}$$

where

$DRMC$ is the Doppler radar's maintenance cost
Y is the number of years in operation

The natural logarithm of the doppler radar's annual maintenance cost, C_{adm}, is expressed as follows [38]:

$$\ln C_{adm} = A_2 + B_2 \ln k_{drf} \tag{6.58}$$

where

$A_2 = -1.269$
$B_2 = 0.696$
k_{drf} is the doppler radar first unit cost expressed in 1974 dollars ($\times 10^3$)

6.4.6 Estimating Software Maintenance Cost

The software maintenance cost can be estimated by using the following cost estimating relationship [39]:

$$SMC = 3(C_{mm}) \cdot (I_{ch})/k_d \tag{6.59}$$

where

SMC is the software maintenance cost
C_{mm} is the cost per man-month
I_{ch} is the number of instructions to be changed on monthly basis
k_d is the difficulty constant

The values of the difficulty constant, k_d, for easy, medium, and hard programs are 500, 250, and 100, respectively.

6.4.7 Estimating Computer (Avionics) Maintenance Cost

In this case, the maintenance cost is expressed as follows:

$$CMC = (C_{cam} Y)/1{,}000 \tag{6.60}$$

where

CMC is the computer maintenance cost
Y is the number of years in operation
C_{cam} is the yearly computer maintenance cost per unit expressed in 1974 dollars ($\times 10^3$)

The natural logarithm of C_{cam} is given in [40] by

$$\ln C_{cam} = A_3 + B_3 \ln C_u - D \ln(MTBF) \tag{6.61}$$

where

$A_3 = 6.944$
$B_3 = 0.296$

$D = -0.63$
$MTBF$ is the mean time between failures expressed in hours
C_u is the unit cost expressed in 1974 dollars ($\times 10^3$)

6.5 Summary

This chapter discussed manufacturing, quality, and maintenance costs. Each of those areas are presented, separately. Estimation of manufacturing costs for new processes is described. Two methods for estimating manufacturing costs are presented along with the manufacturing progress curve. Estimation of average cost per unit involving learning is considered. Ten mathematical models used in estimation of manufacturing costs are presented. These models are concerned with estimation of direct material cost, single point tools' lives, machining cost, tool cost (associated with a cutting tool that is brazed to the toolholder), and so on.

Four categories of quality costs are presented. These are internal failure, prevention, external failure, and appraisal costs. Distributions of quality cost among its elements are given. Four quality cost related indexes are presented.

The topic of maintenance cost estimation is explored. The typical maintenance cost factors that impact cost of ownership are listed. Equations for estimating annual maintenance cost of various different plants are presented. These plants include cement, coke, pulp and paper, and electrolytic. Models for estimating cost of corrective maintenance, system logistic support, fire control radar's maintenance, doppler radar's maintenance, software maintenance, and computer (avionics) are given.

6.6 Problems

1. What are the cost items which should not be overlooked when estimating manufacturing costs for new processes?
2. How would you estimate manufacturing cost in the absence of company records?
3. Discuss the manufacturing cost estimation method called the 1-3-9 rule.
4. What is the learning curve?

5. Assume that the first of a lot of 25 items takes 100 hours to manufacture. If the improvement rate for doubling production is 70%, compute the time to produce a tenth item.
6. What are the major categories of quality costs? Discuss in detail each one of them.
7. What are the important elements of the appraisal cost?
8. What are the advantages and disadvantages of the index used for evaluating vendor quality cost performance?
9. List the typical maintenance cost factors that impact cost of ownership.
10. Assume that the yearly scheduled operating hours for a system are 1900 hours. The values of the system mean time between failures (*MTBF*) and mean time to repair (*MTTR*) are 500 hours and 20 hours, respectively. If the cost of maintenance labor is $15 per hour, calculate the annual cost for corrective maintenance.

References

1. J.K. Carson, Quality Costing—A Practical Approach, International Journal of Quality and Reliability Management, Vol. 3, 1986, pp. 54–63.
2. J.D. Breeze, Quality Costs Can be sold, Proceedings of the American Society for Quality Control Technical Conference, 1980, pp. 795–801.
3. Chemical Engineering (CE) Cost File 92, Estimating Manufacturing Costs for New Processes, Chemical Engineering, August 17, 1964, pp. 160–162.
4. J.H. Black, Operating-Cost Estimation, in Cost and Optimization Engineering, edited by F.C. Jelen and J.H. Black, McGraw-Hill Book Company, New York, 1983, pp. 382–406.
5. H.F. Rondeau, The 1-3-9 Rule for Product Cost Estimation, Machine Design, August 21, 1975, pp. 30–37.
6. E.M. Malstrom, Quantitative Cost Control Methods, in Manufacturing Cost Engineering Handbook, edited by E.M. Malstrom, Marcel Dekker, Inc., New York, 1984, pp. 163–183.
7. T.P. Wright, Factors Affecting the Cost of Airplanes, Journal of Aeronautical Science, Vol. 3, February 1936.
8. E.M. Malstrom, R.L. Shell, A Review of Product Improvement Curves, Manufacturing Engineering, Vol. 82, No. 5, May 1979.
9. E.M. Malstrom, Product Improvement Curves, in Manufacturing Cost Engineering Handbook, edited by E.M. Malstrom, Marcel Dekker, Inc., New York, 1984, pp. 335–354.
10. P.F. Ostwald, Cost Estimating for Engineering and Management, Prentice-Hall, Inc., Englewood Cliffs, New Jersey, 1974.

11. L.M. Matthews, Estimating Manufacturing Costs, McGraw-Hill Book Company, New York, 1983.
12. G.J. Thuesen, W.J. Fabrycky, Engineering Economy, Prentice-Hall, Inc., Englewood Cliffs, New Jersey, 07632, 1984, pp. 421–422.
13. W.J. Fabrycky, P.M. Ghare, P.E. Torgersen, Industrial Operations Research, Prentice-Hall, Inc., Englewood Cliffs, New Jersey, 1972.
14. E.M. Malstrom, Making the Initial Estimate, in Manufacturing Cost Engineering Handbook, edited by E.M. Malstrom, Marcel Dekker, Inc., New York, 1984, pp. 87–109.
15. G.E. Dieter, Engineering Design: A Materials and Processing Approach, McGraw-Hill Book Company, New York, 1983.
16. G. Boothroyd, Fundamentals of Metal Machining and Machine Tools, McGraw-Hill Book Company, New York, 1975.
17. E.J.A. Armarego, R.H. Brown, The Machining of Metals, Prentice-Hall, Inc., Englewood Cliffs, New Jersey, 1969.
18. J. Ham, Group Technology, in Handbook of Industrial Engineering, edited by G. Salvendy, John Wiley & Sons, New York, 1982, pp. 7.8.1.–7.8.19.
19. E.P. Degarmo, Materials and Processes in Manufacturing, Macmillan, New York, 1968.
20. B.W. Niebel, A.B. Draper, Product Design and Process Engineering, McGraw-Hill Book Company, New York, 1974.
21. J.J. Plunkett, B.G. Dale, R.W. Tyrrell, Quality Costs, Department of Trade and Industry, London, U.K., 1985.
22. J.M. Juran, The Economics of Quality, in Quality Control Handbook, edited by J.M. Juran, McGraw-Hill Book Company, New York, 1951, Chapter 1.
23. J. Sittig, The Economic Choice of Sampling Systems in Acceptance Sampling, Bulletin of International Statistics Institute, 1951, pp. 10–30.
24. A.V. Feigenbaum, Total Quality Control, McGraw-Hill Book Company, New York, 1961.
25. H.L. Freeman, How to Put Quality Costs to Use, Transactions of the American Society for Quality Control (ASQC) Metropolitan Conference, 1960, pp. 15–18.
26. J. Campanella, F.J. Corcoran, Principles of Quality Costs, Proceedings of the American Society for Quality Control (ASQC) Annual Conference, 1982, pp. 124–137.
27. Bibliography of Articles Relating to Quality Cost Concepts and Improvement, American Society for Quality Control (ASQC) Quality Cost Technical Committee, January 1981. Available from the American Society for Quality Control, 230 West Wells Street, Milwaukee, Wisconsin 53203.
28. Guide for Managing Vendor Quality Costs, Prepared by the American Society for Quality Control, January 1980. Available from the American

Society for Quality Control, 230 West Wells Street, Milwaukee, Wisconsin 53203.
29. D.H. Besterfield, Quality Control, Prentice-Hall, Inc., Englewood Cliffs, New Jersey 07632, 1979, pp. 252–265.
30. H.J. Harrington, Poor-Quality Cost, Marcel Dekker, Inc., New York, 1987.
31. R.J. Williams, Guide for Reducing Quality Costs, Proceedings of the American Society for Quality Control Annual Conference, 1982, pp. 360–366.
32. L.J. Esterby, Quality Cost Analysis: A Productivity Measure, Proceedings of the American Society for Quality Control Annual Conference, 1981, pp. 432–439.
33. Engineering Design Handbook: Maintenance Engineering Techniques, Document No. AMCP 706-132, 1975, Prepared by Headquarters, United States Army Material Command, 5001 Eisenhower Ave., Alexandria, VA 22333.
34. M.D. Robbins, Maintenance Cost Estimation—Is This a Better Tool for You? Chemical Engineering, Feb. 9, 1959, pp. 140–142.
35. M.D. Robbins, More Light on—Predicting Your Maintenance Costs, Chemical Engineering, July 13, 1959, pp. 172–178.
36. J.W. Hackney, Estimate Production Costs Quickly, in Modern Cost-Engineering Techniques, edited by H. Popper, McGraw-Hill Book Company, New York, 1970, pp. 245–250.
37. D.R. Earles, LCC-Commercial Application Ten Years of Life Cycle Costing, Proceedings of the Annual Reliability and Maintainability Symposium, 1975, pp. 74–85.
38. Cost Analysis of Avionics Equipment, Vol. 1, Prepared by the Air Force Systems Command, Wright-Patterson Air Force Base, Ohio, February 1974. The NTIS Report No. AD 781132, Available from the National Technical Information Service (NTIS), Springfield, Virginia, USA.
39. M.R. Sheldon, Life Cycle Costing: A Better Method of Government Procurement, Westview Press, Boulder, Colorado, 1979.
40. M.E. Earles, Factors, Formulas, and Structures for Life Cycle Costing, Published by Eddins-Earles, 89 Lee Drive, Concord, Mass. 91742, 1978.

CHAPTER 7

Warranties

7.1 Introduction

Nowadays, manufacturers usually provide buyers with written statements guaranteeing the integrity of products they manufacture. The responsibilities of the manufacturers are outlined by these statements in situations when the product happened to be defective. There are several reasons for manufacturers to provide warranties on their industrial products. According to a survey of 369 producers of industrial products [1], the primary reasons were to support company products, to define liabilities and responsibilities specifically, to meet competitive need, etc. This survey has also revealed that the average cost for servicing warranty claims is below 2 percent of a product's selling price. And servicing and repairing of products under warranty may account from a fraction of 1 percent to more than 10 percent of the net sales of a manufacturer. The sales volumes of manufacturers studied varied from under $10 million to over $100 million. In addition, the companies surveyed manufactured products such as electrical and communications products, scientific equipment and control products, and business and commercial equipment.

The United States government attention on warranties and guarantees has been increasing ever since the early seventies. In 1982, the Product Performance Agreement Center (PPAC) was established by the United States Air Force. The purpose of this center is to provide expertise for analyzing, selecting, and developing product performance agreements to Government and industry. In 1984, section 794 of the FY 84 Military Appropriations Act prohibited the procurement of weapons systems, unless, the military contractors provide Government the written guarantees. This chapter presents several different aspects of warranties.

7.2 Factors for Having a Warranty and General Warranty Obligations

There are several factors for having a warranty. Some of those are listed below [2]:

i) To place responsibility of correcting defects in manufactured products on the shoulders of producers
ii) To facilitate manufactured product's quick acceptance
iii) To assure the buyer that the purchased item will at least satisfy the contract requirements
iv) To enhance the product's marketability
v) To encourage the producer to manufacture better products
vi) To encourage better monitoring of performance of products
vii) To expedite payment for the sold product
viii) To allow buyer time to master operating of the product during warranty period with no charge
ix) To pinpoint remedies within user's reach in situation when the purchased item does not satisfy warranty requirements
x) To allow user the time to make decision regarding how and where the purchased product will be maintained once the warranty period is over
xi) To give better assurances that at the moment of need the back-up products and services will be available

In general there are a number of options opened to manufacturers for fulfilling warranty obligations. In particular, the manufacturers of electrical or mechanical products have four basic options as below:

i) Replace faulty components or compensate the buyer for their purchase
ii) Replace the entire faulty product
iii) Make payments for the installation (service labor) of new replacement components, or repair the faulty ones
iv) Make payments for the cost of replacement components or parts and for the labor required to install or repair parts or components in question

According to the survey mentioned earlier most of the manufacturers during the product's warranty coverage period make payments for both replacement of faulty parts and service labor. Normally, the warranty coverage period is for twelve months.

7.3 Basic Elements of a Warranty and Factors for Specifying Warranty Elements

The past experience has shown that the warranties may vary significantly in length and construction depending on several factors. However, in the opinion of many warranty experts there are several primary elements which should be incorporated in almost every warranty. Some of those basic elements are as follows [1]:

i) Warranty duration
ii) Warranty exceptions and exclusions
iii) Warrantor's clear identification
iv) Steps for filing a warranty claim by the purchaser
v) The particular class of purchasers for whom products are warranted
vi) The extent of the seller's responsibilities
vii) The particular parts or components of the product covered under the warranty
viii) Procedures for arbitration for setting warranty disputes, if applicable
ix) Required conditions of product use and maintenance (from the purchaser) if the warranty is to be applicable

There are several factors which should be taken into consideration in specifying elements of warranty. From an airline angle some of those factors are as follows [2]:

i) Financial health of the supplier
ii) Maintenance accessibility
iii) Product line's planned longevity and complexity
iv) State of the art/design
v) Maturity with product's supplier
vi) Warranty element's measurability
vii) Product's producibility
viii) Technical support capability of the supplier
ix) Supplier organization's management capability

7.4 Warranty Administration, and Warranty Service and Repair Costs

Warranty administration is an important function which six-out-of-ten manufacturers assign to their product or technical service. Table 7.1 provides breakdowns of responsibility for warranty administration in manufacturer's organization [1].

There are numerous product and marketing factors which are responsible for the positioning of warranty responsibilities in the company structure. Among the primary factors are those given below:

i) The organizational structure of the company
ii) Frequency of warranty claims
iii) Firm's distribution system characteristics
iv) The importance buyers attach to warranty protection issues
v) The types of warranty claims

Normally, a manufacturer is likely to assign warranty administration responsibility to a sales unit or customer service when the warranty claims are relatively straight forward or small in number.

The sales or customer service unit is also likely to be responsible for warranty administration when the product is sold through multiple channels of distribution. In situations, when claims require significant technical input, the manufacturers are more likely to place it (i.e. warranty administration responsibility) in a product service or technical service unit. A separate warranty administration unit may be created by the company when the volume or cost of claims is rather high.

Table 7.1. Breakdowns of responsibility for warranty administration in manufacturers' organizations.

No.	Unit responsibility for warranty administration	Percentage of manufacturers surveyed [1]
1	Customer service	24%
2	Product or technical service	60%
3	Separate warranty administration unit	10%
4	Others	6%

The warranty servicing cost as a percentage of net sales in roughly 80% of the manufacturers is less than 3%. However, this figure may vary from one type of manufacturer (say, business equipment) to another type (e.g., electrical equipment). The warranty costs are dependent on the length of the warranty period.

In addition, the size of the firm also plays a role in variations of warranty costs. For example, companies with less than $10 million yearly sales tend to spend less on warranties than the larger ones. This fact can be verified from the surveyed data presented in Reference [1].

With respect to keeping records of warranty costs most companies possess such records [1]. Some of the reasons for this are as follows:

i) Usefulness of warranty costs for measuring product performance
ii) Providing input to product pricing formulas so that the costs of warranty are fully recovered
iii) Natural concern (of warranty costs) in service planning

7.5 Reasons for the Growing Warranty Claims and Future Warranty Problems

Many manufacturers have reported a growing number of warranty claims by the buyers. There could be numerous reasons for their growth. Some of those are listed below:

i) Poor product design
ii) Increase in customer awareness of producers' warranty obligations and their (customers) warranty rights
iii) Increase in misuse of products by customers
iv) Degradation in quality of products
v) Inflationary pressures, generating greater sensitivity to costs of repairs
vi) Increase in complexity of products
vii) More service-sensitive products
viii) Lack of experienced and qualified maintenance manpower in the buyers' organizations

The service executives of many manufacturing companies have forecasted numerous warranty related problems in the future. Some of those forecasted problems are as follows [3]:

i) Increase in user expectations
ii) Rapid increase in labor cost, cost of materials, and so on

iii) Educating product dealers and users regarding use of products
iv) Government regulations
v) Training and hiring competent people for product service
vi) Increase in product liability activity
vii) Increase in litigation
viii) Keeping hold on competent service personnel
ix) Maintaining warranty starting dates' records

7.6 Long Term Warranties

Generally, the long term warranty is required by customers when the equipment capital cost is high. Aircrafts are the one example of high capital cost equipment. The warranty coverage for commercial airplanes such as the DC-10 is for as long as 10 years or 30,000 flying hours. The Douglas Aircraft Company covers items such as the following in their "standard warranty" for aircraft (e.g., DC-10s) [4]:

i) Defects found in installation of vendor components (DC-10 warranty coverage: parts not subject to prior inspection are warranted for 5,000 flying hours or 2 years/8,000 flying hours or 3 years)
ii) Service life policy for primary structure of the aircraft (DC-10 warranty coverage: 30,000 flying hours or 10 years)
iii) Faults in workmanship and in material (DC-10 warranty coverage: parts not subject to prior inspection including primary structure are warranted for 3,000 flying hours or 3 years/5,000 flying hours or 2 years)
iv) Guarantees for performance (DC-10 warranty coverage: 400 flying hours or 3 months)
v) Service policy for basic landing gear parts (DC-10 warranty coverage: 30,000 flying hours or 10 years or 20,000 landings)
vi) Costs for transportation (DC-10 warranty coverage: one way paid by the Douglas Aircraft Company)
vii) Faults found in design (DC-10 warranty coverage: 1.5 years)
viii) Reimbursement of warranted labor (DC-10 warranty coverage: labor imbursed with the exception of inspection checkout and test effort)
ix) Faults discovered in selection of materials (DC-10 warranty coverage: 1.5 years)

x) Recycling of warranted period for replacement, redesign or repair (DC-10 warranty coverage: yes)
xi) Negligence in conforming to the detail specification (DC-10 warranty coverage: parts not subject to prior inspection including primary structure are warranted for 3 years or 8,000 flying hours/2 years or 5,000 flying hours)
xii) Faults discovered in the process of manufacture (DC-10 warranty coverage: 1.5 years)
xiii) Making reimbursement payments at the direct labor rate of the operator, as well as, 50% for burden (DC-10 warranty coverage: yes).

7.7 Selecting Equipment for Warranty Coverage and Benefits of Warranty to Manufacturer and User

It is not wise to select equipment indiscriminately for warranty coverage. There are several factors which must be considered carefully in selecting equipment for such coverage. Some of those factors [5] are field-testability of the equipment; initial support cost (moderate to high); and knowledge about the equipment application: expected operating time, use environment, self-containability of the equipment, transportability of the equipment, and the development of the equipment (to a level where its expected reliability and maintainability can be estimated reasonably accurately).

Both the equipment manufacturer and the user benefit from warranties. Some of their primary benefits are shown in Figure 7.1.

7.8 Ways for Obtaining Greater Benefits under Warranties and the Ideal Warranty Contract

There are various ways for obtaining greater benefits under warranties. The General Accounting Office (GAO) identified five principal ways in a report to the Congress, issued in 1975 [6]. Four of those are as follows:

i) Upgrade management surveillance
ii) Acquire improved warranty terms
iii) Improve warranty related information for users of equipment
iv) Increase use of billback agreements

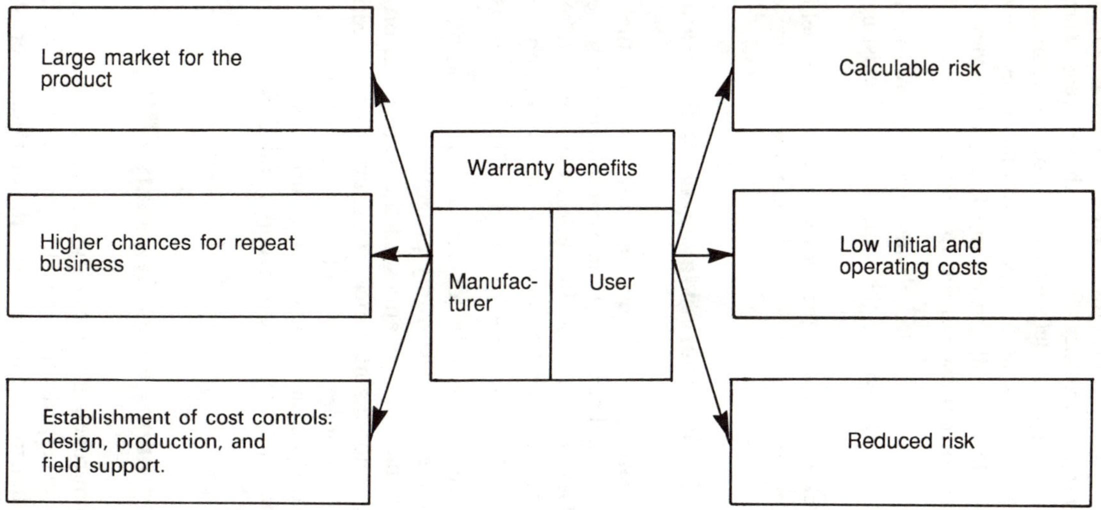

Figure 7.1. Warranty benefits to manufacturer and user.

In 1975, the Army Procurement Research Office (APRO) issued a report [7, 8] regarding warranties. This report identified areas in which field procurement and other functional area personnel lack guidance concerning warranties. Some of those areas were type of warranty provision, type of defects, evaluation of commercial warranties, and legal implications of type of technical description.

According to Reference [8], presenting U.S. Army viewpoints, an ideal warranty contract possesses the following:

i) Incorporates warranty related claims reporting by the user
ii) Functions effectively within the framework of existing support system
iii) Provides quick response to the requirements of the field user
iv) Possesses monetary incentives for the equipment user

7.9 Assessing the Warranty Management Program

This section presents a list of questions for assessing the warranty management program. These questions were developed for each U.S. Army Material Development and Readiness Command (DARCOM) major subordinate commander to assess the command's warranty management program. The questions are presented below [8]:

i) What equipment is currently covered by the warranty?
ii) Who is in-charge of the warranty program?
iii) What is the impact of warranties on the acquisition process?
iv) How supply and maintenance programs for peacetime and planned wartime operations affected by warranties?
v) What policy do you have on warranty utilization?
vi) What are the equipment/parts recommended for the warranty consideration?

7.10 Reliability Improvement Warranty (*RIW*)

The reliability improvement warranty is one of the procurement methods and is developed under the life cycle cost policy umbrella. Under reliability improvement warranty, the manufacturer of product is agreeable to correct all covered failures at no extra charge for a stated time. The fundamental objective of reliability improvement (warranty) is to improve equipment reliability and cut down repair costs. In the United States Department of Defense the concept of reliability improvement warranty was first put forward in a memorandum [9] to the assistant secretaries of the Military Departments [10], in 1974. The memorandum entitled "Trial Use of Reliability Improvement Warranties in the Acquisition Process of Electronic Systems/ Equipment-ACTION-MEMORANDUM" called for the trial use of reliability improvement warranty (*RIW*) in the procurement process to help determine its scope and benefits.

In article [11] published in the Proceedings of the 1978 Annual Reliability and Maintainability Symposium reported that the United States Air Force has committed $65.8 million to *RIW* and this figure is expected to grow in the future. The distribution of the committed *RIW* amount for various projects is as follows:

i) Project A (F-16 aircraft components: $44 million)
ii) Project B (ARN-118 TACAN (a short-range navigation system): $14 million
iii) Project C(C-141/KC-135 Inertial Navigation System: $3.3 million)
iv) Project D (C-141 Altitude Heading and Reference System (AHRS): $2.2 million)
v) Project E (C-130 Omega Navigation Set: $1.8 million)
vi) All other projects: $0.5 million

The items included in Project A were radar computer, flight control computer, radar transmitter, hud electronics, radar antenna, heads up display, radar low power RF, navigation unit, and radar digital signal processor. The items under "All other Projects" were Klystron electron tubes, F-111 displacement gyro, C-130 hydraulic pump, and AVU-8 C/A airspeed indicator.

7.10.1 Analyses for Reliability Improvement Warranty

This section presents four types of analyses required by the United States Air Force (USAF) prior to applying reliability improvement warranty (*RIW*). The types of analyses required are shown in Figure 7.2 [11]. Those are appraisal of *RIW* needs in request for proposal, *RIW* proposals evaluation, equipment/program screening, and cost analysis of *RIW*/*NON-RIW*. The purpose of appraisal of *RIW* needs in request for proposal is threefold: a) assure continuity of *RIW* terms with technical specifications, b) assure balanced risks/benefits, and c) assure realistic government/contractor duties. The second type

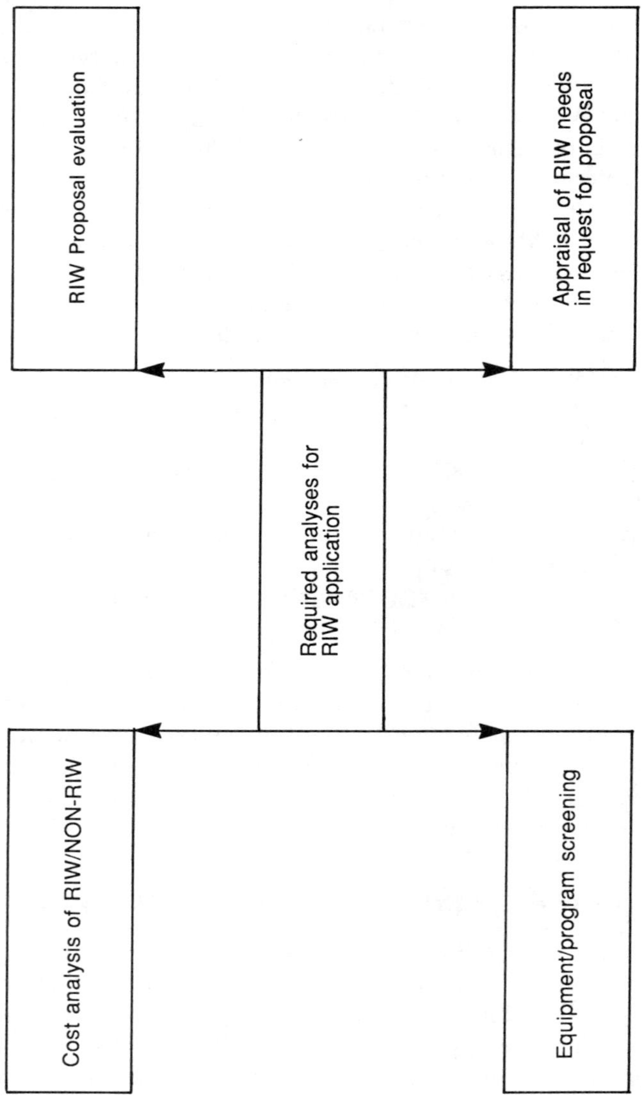

Figure 7.2. Types of analyses required for *RIW* application.

of analyses (i.e., *RIW* proposals evaluation) has the following two purposes:

i) Make verification of reliability realism
ii) Determine positive and negative incentives' balance

The sole purpose of equipment/program screening is to identify those programs/projects which have significant potential for *RIW* benefits. The fourth type of analyses (i.e., cost analysis of *RIW/NON-RIW*) has one specific objective, i.e., to establish cost effectiveness of reliability improvement warranty.

7.10.2 Concerns Associated with Reliability Improvement Warranty

The USAF Air Staff and the logistics community in general [11] voiced several concerns with the reliability improvement warranty concept. Those concerns range from what will happen upon termination of the *RIW* to issues associated with funding and payment. A list of selected concerns is presented below [11]:

i) Warranty parameters' measurement
ii) Using acquisition funds instead of operation and maintenance funds for purchasing warranty
iii) Control of configuration under reliability improvement warranty
iv) Sufficiency of required technical data
v) Sufficiency of required training
vi) Making payments for the warranty price ahead of processing of failed items
vii) Wartime surge needs for items under reliability improvement warranty
viii) Performance's (guaranteed mean time between failures) commercial impossibility
ix) Impact of delayed installation of items under warranty.

7.10.3 Approaches to Warranties or Incentives

The advanced Medium Range Air-to-Air Missile (AMRAAM) program office of the United States Air Force formed a Warranty/Incentive Working Group (WIWG), in December 1979; to examine approaches and develop a strategy for implementing warranties and incentives on the program. The members of the WIWG evaluated several approaches [12] to warranties or incentives. They ranged from reliability improvement warranty to verification guarantee. Some of the approaches considered were as follows:

i) Mean time to repair guarantee
ii) Guaranteed mean time between failures
iii) Verification guarantee
iv) Guaranteed availability
v) Technical data warranty
vi) Reliability improvement warranty
vii) Basic commercial warranty
viii) Reliability incentive fee
ix) Supply warranty

A list of sixteen candidate warranty types is presented in Reference [12]. The WIWG concluded that the following approaches to warranty are feasible for the Advanced Medium Range Air-to-Air Missile:

i) Reliability improvement warranty
ii) Reliability incentive fee
iii) Guaranteed availability
iv) Verification guarantee

7.10.4 Conditions Needed for the Implementation of an Effective Reliability and Maintainability Field Warranty

For the implementation of an effective reliability and maintainability field warranty, the numerous conditions are required. A number of them is presented below [13]:

i) The program requires no additional personnel and is quite straightforward to administer
ii) Reliability and maintainability (R&M) evaluation tests have been performed on the equipment under warranty
iii) The corrective measures indicated by the R&M evaluation tests have been implemented
iv) The equipment under warranty are self-contained items
v) The equipment under warranty is repaired timely and the contractor supplies correct number of spares
vi) The contractor possesses knowledge of the field utilization
vii) The equipment inherits fault isolation capabilities

7.10.5 Data Items Required for Administering and Supporting the Reliability Improvement Warranty

This section presents data items which were required for administering and supporting the Army Lightweight Doppler Navigation System (LDNS) [14] reliability improvement warranty. Figure 7.3 shows the required items. These are a data collection and analysis plan, a semi-annual warranty data report, and an annual warranty effectiveness study report. The first data item, i.e. the data collection and analysis plan, outlines the way the contractor will collect, process, analyze, and report the needed information. The second data item, i.e., semi-annual warranty data report, possesses the records, as below:

i) Summaries pertaining to corrective-measures
ii) Summary pertaining to labor/material usage
iii) Delivery and shipment dates of set/unit
iv) Warranty population
v) Item failure experience analysis

The third data item, i.e., the annual warranty effectiveness study report, presents experiences of the contractor, as well as, conclusions concerning the warranty concept's effectiveness.

7.10.6 Reliability Improvement Warranty with a Mean Time between Failures Guarantee

In situations where the purpose is to achieve major reliability growth of systems, the mean time between failures (*MTBF*) guarantee can be used [15]. Over consecutive periods of the guarantee an increasing series of target values is stated. The contractor corrective action results, when the target values are not met. On the other hand the incentive awards could result when the targets are exceeded.

7.10.7 Manufacturer Reliability Improvement Warranty Risk Factors

In order to design and manufacture reliable equipment all reliability improvement warranty contracts include incentives for manufacturers. Even though there are several methods for structuring terms and conditions associated with reliability improvement warranty (*RIW*) to take care of contingencies, the contractors are still faced with *RIW*

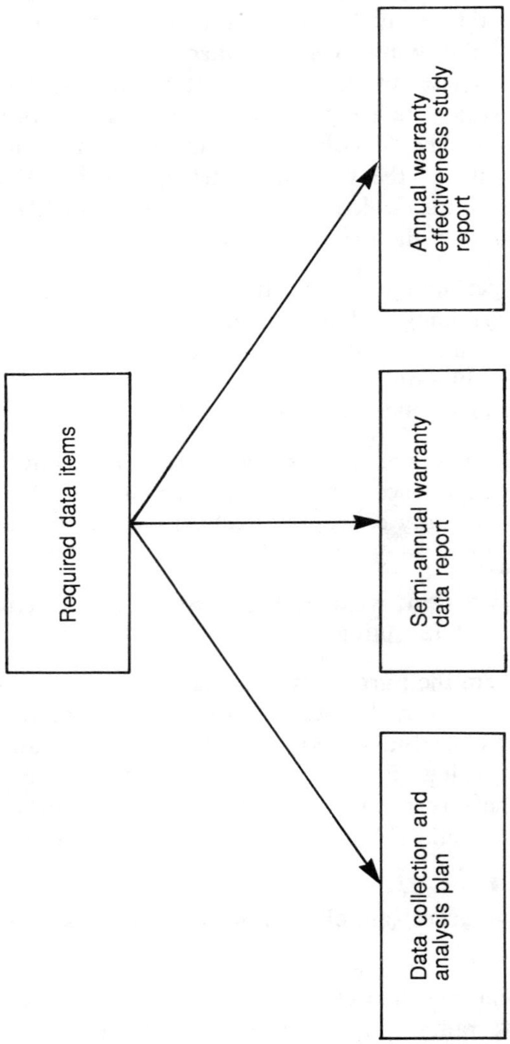

Figure 7.3. Data items required for administering and supporting the reliability improvement warranty.

risk factors. Some of the manufacturer *RIW* risk factors are as follows [16]:

i) Usage rate of the equipment is not exactly known
ii) Intention to use warranty is made known to the manufacturer at a late stage
iii) Uncertainties in reliability–prediction
iv) Using the reliability improvement warranty to advanced technology
v) Failures beyond the control of the manufacturer
vi) Substantial number of unverified failures sent back to manufacturer
vii) Procedures for engineering change proposals' approval are rather time-consuming
viii) During bidding destinations for shipping of items under warranty may not be known
ix) Inflation rates' unpredictability for long periods
x) Effect of events beyond the control of the manufacturer on turnaround time. An example of such events is strikes.

For all the above risk factors, the risk reduction approaches are given in Reference [16].

7.11 Warranty Related Cost Models

This section presents a number of selected warranty related cost models.

Model I

This model is concerned with pricing the basic reliability improvement warranty (*RIW*). The manufacturer pricing model for the basic *RIW* is expressed below [17].

$$FPMW = MFC + C_{IA} + PF + CD + F \tag{7.1}$$

where

$FPMW$ is the fixed price paid for the warranty to the manufacturer
MFC is the manufacturer's fixed cost associated with warranty
C_{IA} is the cost associated with improvement actions to attain $MTBF_X$, the achieved mean time between failures (reliability improvement warranty period average)

PF is the profit
CD is the damages associated with not meeting the required turnaround time

The cost, F, is defined by

$$F = (CCT)(N)(WP)(UR)/MTBF_X \tag{7.2}$$

where

CCT is the manufacturer's cost per unit repair
N is the quantity of systems to be delivered
WP is the length of the warranty period
UR is the rate of usage expressed in operating time per calendar time
$MTBF_X$ is the achieved mean time between failures (reliability improvement warranty period average)

The following formula was used in a reliability improvement contract to estimate liquidated damages, CD, at each measurement point:

$$CD = 25(TAT - TATC)(RU) \text{ dollars}, \quad TAT > TATC \tag{7.3}$$

where

TAT is the mean measured turnaround time
RU is the quantity of units returned in the measurement period
$TATC$ is the ceiling of the turnaround time

Model II

This model is concerned with estimating contractor warranty cost, CWC, over the warranty time period. The cost, CWC, is given by [16]:

$$CWC = (MCR) \cdot (TO) \cdot (FR) + CFC \tag{7.4}$$

where

MCR is the contractor's mean cost for repairing a unit sent back for warranty service
TO is the total number of operating hours of all equipment under warranty during the warranty period
FR is the mean failure rate per hour of all equipment under warranty during the warranty period
CFC is the fixed cost to the contractor associated with the warranty

Let us now examine risks associated with warranty pricing from the point of view of return on sales versus mean time between failures ($MTBF$) variation. The return on sales is defined by

$$RT_S = \frac{PF}{P_{RIW}} \tag{7.5}$$

where

RT_S is the return on sales
PF is the profit
P_{RIW} is the reliability improvement warranty price

The ratio of the actual mean time between failures ($MTBF_{AC}$) to average mean time between failures ($MTBF_{BD}$) at which the warranty was bid is related to the warranty pricing variables as follows:

$$\frac{MTBF_{AC}}{MTBF_{BD}} = \frac{(MCR) \cdot (TO)}{Z} \tag{7.6}$$

where

$$Z = F_L(1 - RT_S)\{(MCR) \cdot (TO) + (MTBF_{BD})(CFC)\} - (CFC)(MTBF_{BD}) \tag{7.7}$$

The symbol F_L represents loading applied to expected costs to produce profit. Normally, the value of this factor (F_L) will be greater than unity. At the time of pricing, the contractor's perception of his risks may very well influence the value of F_L.

For the fixed value of RT_S, the lower the ratio of $MTBF_{AC}$ to $MTBF_{BD}$, the better for the contractor. It can easily be seen from Equation (7.7) that as the value of F_L decreases the ratio of $MTBF_{AC}$ to $MTBF_{BD}$ goes up. Some plots of F_L versus ratio of $MTBF_{AC}$ to $MTBF_{BD}$ are given in Reference [16].

Model III

This model is known as the warranty price model and is based on Equation (7.4). The warranty price is given by

$$P_W = F_L[(MCR_{BD})(TO_{BD})(FR_{BD}) - CFC_{BD}] \tag{7.6}$$

The subscript BD in the above equation signifies a "bid" value for pricing and P_W is the warranty price. The factor F_L and other symbols are defined in Model II.

Model IV

This model is concerned with pricing the reliability improvement warranty with mean time between failures (*MTBF*) guarantee. In this case, the contractor or manufacturer provides guarantees that the product will satisfy stated *MTBF*s by given points in time. If the specified *MTBF* is not met by the stated point in time, the contractor provides consignment spares according to an agreed formula. The fixed price paid for the warranty to the manufacturer [17] is

$$FPMW = MFC + C_{IA} + PF + CD + F + (CSC)(NCS) \quad (7.7)$$

where

NCS is the quantity of consignment spares needed
CSC is the cost per consignment spare

The symbols *FPMW*, *MFC*, C_{IA}, *PF*, *CD*, and *F* are the same as for the Model I. A typical formula for determining consignment spares is as follows:

$$NCS = \left[\frac{\text{Guaranteed mean time between failures}}{\text{Measured (achieved) mean time between failures}} - 1\right](TSL) \quad (7.8)$$

where *TSL* is the target spares level

Model V

This model is also associated with reliability improvement warranty and can be used to estimate cost of the contractor, CCT_R, for repairing units sent back for an identified failure mode. Thus, the contractor cost is expressed [17] by

$$CCT_R = (RC + HC)(UT)(FR) \int_{T_0}^{T_W} S_q(x)\, \mathrm{d}x \quad (7.9)$$

where

T_W is the end time of warranty
RC is the cost of repair per failure of identified failure mode
UT is the utilization in mean operating time per time per system
HC is the cost of handling per unit sent back (e.g. packing, failure identification)
$S_q(x)$ is the number of systems delivered at time x
FR is the identified failure mode's failure rate

Model VI

This model is concerned with estimating life cycle cost, with warranty. The life cycle cost, C_{LCW}, with warranty [5] is defined by

$$C_{LCW} = (RSC_W)(CT) + (UP)(M_W)(AF) + DMC_W + A \quad (7.10)$$

and

$$A = (1 + CF)(DWRC_W + MDC_W)RD(CT) \quad (7.11)$$

The subscript W in above two equations denote a warranty procurement. The remaining main symbols are defined below:

C_{LCW} is the life-cycle cost for a procurement with warranty over interval $(0, CT)$
RSC_W is the recurring support costs per month
CT is the calendar time expressed in units of months
UP is the unit's procurement price
M_W is the quantity of units procured
AF is the amortization factor for the time interval $(0, CT)$ and is given by CT over expected life of the equipment
$DWRC_W$ is the direct warranty repair cost to the contractor
CF is the fee of the contractor
DMC_W is the direct maintenance cost to the user
MDC_W is the cost for modification, amortized, and discounted to the contractor
$RF(CT)$ is the risk factor applied by the contractor to costs for a warranty time period, CT. The value of the risk factor, $RF(CT)$, is estimated from the following relationship:

$$RF(CT) = (1 + i)^{CT/12}, \quad 0 < i < 1 \quad (7.12)$$

An illustrative numerical example for indicating the utility of this type of models is presented in Reference [5].

Model VII

This model is concerned with estimating expected unit warranty cost for an item subject to a specific policy, β. Under policy β the buyer receives full compensation for any failures occurring in time interval $(0, t_1)$ and prorated compensation for those failures occurring in the time interval (t_1, t_2). The manufacturer's cost of a failure is defined [18] by

$$C(Y_j) = \begin{cases} C_f, & Y_j < t_1 \\ C_f\left[\dfrac{t_2 - Y_j}{t_2 - t_1}\right], & t_1 \leq Y_j \leq t_2 \\ 0, & \text{otherwise} \end{cases} \tag{7.13}$$

where

Y_j is the time between failures j and $j-1$; $Y_j > 0$
$C(Y_j)$ is the manufacturer's cost of failure j
C_j is the unit's replacement cost
t_1 is the full compensation time limit
t_2 is the prorata compensation time limit

Assumptions such as below are associated with this model:

i) Failures are statistically independent and identically distributed
ii) At each failure the item is replaced
iii) The replacement item has the original warranty

A stochastic process $\{AFC(x)\colon 0 < x < X\}$ describes the total unit warranty cost accumulated for a certain item in a length of warranty cycle X. The accumulated failure cost at time x is given by

$$AFC(x) = \sum_{j=1}^{NF(x)} C(Y_j), \quad x < X \tag{7.14}$$

where $NF(x)$ is the number of failures in time interval $(0, x)$. The expected unit warranty cost for an item is given by

$$C(t_1, t_2) = E\left\{E\left[\sum_{j=1}^{NF^*} C(Y_j)\right]\right\}$$

$$= E\{NF^* E[C(Y_j)]\} \tag{7.15}$$

$$= \frac{c_f}{[1 - F_Y(t_2)]} \int_{t_1}^{t_2} \left[\frac{F_Y(u)}{(t_2 - t_1)}\right] du \tag{7.16}$$

where

NF^* is the quantity of failures in the warranty cycle
E is the expected value
$F_Y(y)$ is the cumulative distribution function for Y

The detailed derivation of Equation (7.16) is shown in Reference [18].

Example 7.1

Manufacturer of an item provides warranty for a 48-month period. The warranty covers total replacement for failures within the initial 12-months of sale and prorated compensation for the remaining 36-months. The failed unit is replaced at the cost of \$2,000. The failure times of the item are described by an exponential distribution. The item's mean time to failure is 60 months. Thus, the cumulative distribution function associated with the item is

$$F_Y(y) = 1 - e^{-[y/60]} \tag{7.17}$$

Using Equation (7.16) calculate the expected unit warranty cost. By substituting the above given data into Equation (7.16), we get

$$\begin{aligned}
C(t_1, t_2) &= \frac{2{,}000}{1 - \{1 - e^{-[48/60]}\}} \int_{12}^{48} \frac{\{1 - e^{-[y/60]}\}}{48 - 12} dy \\
&= \frac{2{,}000}{(36)e^{-[48/60]}} \int_{12}^{48} \{1 - e^{-[y/60]}\} dy \\
&= \frac{2{,}000}{16.1758} [y + 60e^{-y/60}]_{12}^{48} \\
&= \frac{2{,}000}{16.1758} (13.8359) \\
&= \$1710.69
\end{aligned}$$

Thus, the expected unit warranty cost is \$1710.69.

7.12 Summary

This chapter presented several different aspects of warranties. Factors for the warranty need and general warranty obligations are discussed. Nine basic elements of a warranty are listed along with factors for specifying warranty elements. The topic of warranty administration is discussed and the primary product and marketing factors responsible for the positioning of warranty responsibilities in the company structure are given. The warranty service and repair costs are explained. Numerous reasons for the growing warranty claims are presented along with future warranty problems. Long term warranties associated with aircrafts are discussed. The topic of selecting equipment for the warranty coverage is presented. Benefits of warranties to both manufacturers and users are given. A list of questions for assessing the warranty management program is presented.

The subject of reliability improvement warranty (*RIW*) is described in detail. The types of analyses required for the (*RIW*) application are presented and the concerns associated with *RIW* are listed. A number of approaches to warranties or incentives are given along with conditions required for implementation of an effective reliability and maintainability field warranty. Data items required for administering and supporting *RIW*, as well as, manufacturer *RIW* risk factors are presented. The topic of *RIW* with a mean time between failures guarantee is discussed. Seven warranty related cost models are examined.

7.13 Problems

1. Discuss the background of reliability improvement warranty.
2. What are the reasons for having a warranty?
3. Discuss the basic elements of a warranty?
4. List at least eight reasons for the growth of warranty claims.
5. What factors would you consider in selecting equipment for the warranty coverage?
6. What are the advantages and disadvantages to manufacturer and user of having a warranty?
7. What questions would you ask in assessing the warranty management program?
8. What are the concerns associated with *RIW*?
9. List contractor *RIW* risk factors?
10. What are the conditions required for the implementation of an effective reliability and maintainability field warranty?
11. Assume that a manufacturer of a product provides warranty for a 4.5 years period. The warranty covers total replacement for failures within the first 365 days of sale and prorated compensation thereafter. The failed unit is replaced at the total cost of $1,500. The failure times of the product are represented by an exponential distribution with a mean of 65 months. Using Equation (7.16) determine the expected unit warranty cost.

References

1. E.P. McGuire, Industrial Product Warranties: Policies and Practices, A research report from the Conference Board, Conference Board Report

No. 800, 1980. Available from the Conference Board, Inc., 845 Third Avenue, New York, New York, 10022.

2. W.W. Flottman, M.R. Worstell, Mutual Development, Application and Control of Suppliers Warranties in American Airlines, Inc. and Litton Systems-Aero Products Division, Proceedings of the Annual Reliability and Maintainability Symposium, 1977, pp. 213–221.
3. B.S. Dhillon, Engineering Management: Concepts, Procedures and Models, Technomic Publishing Co., Inc., Lancaster, Pennsylvania, 1987.
4. H. Bayer, R.N. Speir, Long Term Commercial Warranty, Proceedings of the Annual Reliability and Maintainability Symposium, 1978, pp. 50–54.
5. H. Balaban, B. Retterer, The Use of Warranties for Defense Avionics Procurement, Proceedings of the Annual Reliability and Maintainability Symposium, 1974, pp. 363–368.
6. "Savings Expected from Better Use of Truck Warranties by Government Agencies", Report to the Congress by the Comptroller General of the United States, General Accounting Office (GAO) Report No. PSAD 75-64, March 20, 1975, Washington, D.C.
7. E.C. Beekler, H. Candy, Analysis of AMC's Use of Warranties, Report No. APRO-507, 1975, US Army Procurement Research Office, (APRO), USALMC, Fort Lee, Virginia.
8. R.A. Mlinarchik, Assuring Army User Satisfaction Through Warranties, Proceedings of the Annual Reliability and Maintainability Symposium, 1980, pp. 307–313.
9. A.T. Mendolia, M.R. Currie, Trial Use of Reliability Improvement Warranties in the Acquisition Process of Electronic Systems/Equipment-Action-MEMORANDUM, August 14, 1974, Department of Defense, Washington, D.C.
10. C.A. Hardy, R.J. Allen, Reliability Improvement Warranty Techniques and Applications, Proceedings of the Annual Reliability and Maintainability Symposium, 1977, pp. 222–228.
11. D.G. Newman, L.D. Nesbitt, USAF Experience with RIW, Proceedings of the Annual Reliability and Maintainability Symposium, 1978, pp. 55–61.
12. P.B. Guglielmoni, An Approach to Selecting Warranties/Incentives, Proceedings of the Annual Reliability and Maintainability Symposium, 1986, pp. 166–170.
13. L. Neoff, Field Operational R&M Warranty, Proceedings of the Annual Reliability and Maintainability Symposium, 1988, pp. 227–230.
14. R. Kowalski, R. White, Reliability Improvement Warranty (RIW) and The Army Lightweight Doppler Navigation System (LDNS), Proceedings of the Annual Reliability and Maintainability Symposium, 1977, pp. 237–241.
15. N.G. Fleig, The United States Air Force Product Performance Agreement Center, Proceedings of the Annual Reliability and Maintainability Symposium, 1986, pp. 449–453.

16. H.S. Balaban, M.A. Meth, Contractor Risk Associated with Reliability Improvement Warranty, Proceedings of the Annual Reliability and Maintainability Symposium, 1978, pp. 123–129.
17. R.K. Gates, R.S. Bicknell, J.E. Bortz, Quantitative Models Used in the RIW Decision Process, Proceedings of the Annual Reliability and Maintainability Symposium, 1977, pp. 229–236.
18. M.U. Thomas, Optimum Warranty Policies for Non-Repairable Items, IEEE Transactions on Reliability, Vol. R-32, 1983, pp. 282–288.

CHAPTER 8

Cost Estimation Models

8.1 Introduction

Over the years, a large number of cost estimation models have evolved in various diverse disciplines, e.g., software engineering, chemical engineering, and telecommunication engineering. In broad terms, a cost model may simply be described as an approach, based on technical and programmatic parameters, for computing concerned costs.

The cost estimation models have many dimensions and their cost elements' values are obtained either from the input data directly or using cost estimating relationships (*CER*s). In some cases, a cost estimating relationship by itself may also be called a cost model. Some of the possible dimensions of a cost model are elements of cost, time, and cost breakdown structure. A number of desirable features are associated with a cost model. Therefore, in designing a model, the factors such as listed below should be considered.

i) Cost to develop, operate, and alter
ii) Feasibility of data requirements
iii) Speed to set up, operate, and change
iv) Tolerance of input errors
v) Operation ease
vi) Inclusiveness and authoritativeness
vii) Capability for sensitivity analyses

There are various types of cost estimation models: life cycle cost models, operation and maintenance cost models, capital cost models, etc. This chapter presents a large number of cost estimation models selected from the published literature. These models may be useful in life cycle cost estimation studies directly or indirectly. Some of the models presented in this chapter, even though they do not directly yield the cost estimate, provide useful inputs for estimating costs. When using some of the cost estimating relationships presented in this chapter, one should pay attention to using the appropriate year for estimated dollars. The year relationship was developed or published may be assumed as the year for estimated dollars (e.g., in 1974 dollars).

8.2 Selected Models

This section presents several models considered to be beneficial in performing cost estimation and related studies.

8.2.1 Cost Estimation Model I

This model is concerned with estimating plant cost and was developed by Cran [1, 2]. The total plant cost is given by

$$\begin{aligned} PC &= (DC) + IDC \\ &= (DC) + (DC)(ICF) \end{aligned} \tag{8.1}$$

where

PC is the plant cost expressed in dollars
DC is the direct cost expressed in dollars
IDC is the indirect cost expressed in dollars
ICF is the indirect cost factor

The direct cost, DC, is expressed by

$$DC = (EC)(DCF_p) + (IC)(DCF_I) \tag{8.2}$$

where

DCF_p is the direct cost factor associated with the plant
DCF_I is the direct cost factor associated with instruments
IC is the cost associated with instruments
EC is the cost associated with equipment

The mean values of the plant direct cost factor, DCF_p, and the instrument direct cost factor, DCF_I, are 2.16 and 2.50, respectively. The indirect cost factor, ICF, can be estimated from the following relationship:

$$ICF = 1.36 - 0.073 \ln DC \tag{8.3}$$

8.2.2 Cost Estimation Model II

This model was developed by Tyszkiewicz [3] for estimating spares' cost, SC_{AS}, for the new aircraft system. The model makes use of the spares cost, SC_{ES}, for an existing aircraft system and adjusts it by a comparison factor, CF, reflecting the differences in system repairability, reliability, hardware complexity, and cost. The new aircraft system's spares cost is expressed by

$$SC_{AS} = (CF)(SC_{ES}) \tag{8.4}$$

The comparison factor, CF, is expressed in terms of the operational and support parameters and can be estimated from the following equations:

$$CF = \alpha(CIR)\left[\frac{I_{MNS}}{I_{MES}}\right] \tag{8.5}$$

$$\alpha = F_1\left[\frac{NFAC_1}{EFAC_1}\right] + F_2\left[\frac{NFAC_2}{EFAC_2}\right] + F_3\left[\frac{NFAC_3}{EFAC_3}\right] \tag{8.6}$$

where

F_i is the ith fraction of the total investment spares "lay-in" value estimated for the existing systems:

$i = 1$ (airframe-related)
$i = 2$ (propulsion-related)
$i = 3$ (equipment-related).

$NFAC_i$ is the ith segment of the "fly-away" cost for the new system:

$i = 1$ (airframe)
$i = 2$ (propulsion)
$i = 3$ (equipment).

$EFAC_i$ is the ith segment of the "fly-away" cost for the existing system:

$i = 1$ (airframe)
$i = 2$ (propulsion)
$i = 3$ (equipment).

I_{MES} is the existing aircraft system's established maintenance index expressed as maintenance man-hours per flying hour

I_{MNS} is the new aircraft system's estimated maintenance index expressed as maintenance man-hours per flying hour

CIR is the quantifier of the cost impact associated with a shift in the category of spares from "depot-repaired" to "base-repaired" or vice-versa between the two aircraft systems in question (i.e., the new and the existing). The value of *CIR* is equal to unity when the change in the ratio of the two categories is zero. In Reference [3], a specific plot is provided to estimate the value of the *CIR*. In addition, an example is presented to demonstrate the application of this model.

8.2.3 Cost Estimation Model III

This model is concerned with estimating reliability acquisition cost, *RAC*, when the state-of-the-art system acquisition cost, SAC_{SA}, and the reliability improvement [mean time between failures (*MTBF*)] ratio, *IR*, compared to state of the art are known. The reliability acquisition cost is given [4] by

$$RAC = 0.2(SAC_{SA}) \ln IR \tag{8.7}$$

8.2.4 Cost Estimation Model IV

A weapon system's cost may be expressed in terms of effective missiles in the system [5]. The effective missiles are the one which intercept and destroy enemy targets successfully. In the event when the weapon system's total cost is apportioned to the manner described above, the cost, *EMC*, per effective missile may be estimated from the following equations [5]:

$$EMC = \frac{TWSC}{(\alpha_{ms}) \cdot X} \tag{8.8}$$

$$X = \frac{NEM}{\alpha_{ms}} \tag{8.9}$$

where

α_{ms} is the quantity of missiles in ready storage
NEM is the quantity of effective missiles
TWSC is the total cost of the weapon system
X is the weapon system effectiveness ratio

This model is described in detail in Reference [5].

8.2.5 Cost Estimation Model V

The model was proposed by Carhart and Herd [6, 7] for estimating the development cost. The model takes into consideration the reliability factor. The development cost is expressed by

$$DC = BC + RC$$
$$= BC + SC_R\left[\frac{MTBF_{ID}}{MTBF_{SD}}\right]^{\beta} \tag{8.10}$$

where

DC is the development cost
BC is the basic cost independent of reliability
RC is the cost dependent on reliability (reliability related cost)
SC_R is the "standard" cost to develop an item having "standard" or current reliability
β is the constant whose value is to be estimated from empirical studies
$MTBF_{SD}$ is the standard design mean time between failures
$MTBF_{ID}$ is the improved design mean time between failures

Let us now assume that the reliabilities of the standard design and of the improved design, respectively, are given by

$$R_{SD}(t) = e^{-t/MTBF_{SD}} \tag{8.11}$$

and

$$R_{ID}(t) = e^{-[t/MTBF_{ID}]} \tag{8.12}$$

where

$R_{SD}(t)$ is the reliability of the standard design
$R_{ID}(t)$ is the reliability of the improved design
t is time

Taking natural logarithms of Equation (8.11) yields

$$\ln R_{SD}(t) = -\left[\frac{t}{MTBF_{SD}}\right] \tag{8.13}$$

Rearranging Equation (8.13) results in

$$MTBF_{SD} = -\left[\frac{t}{\ln R_{SD}(t)}\right] \tag{8.14}$$

Similarly, we get

$$MTBF_{ID} = -\left[\frac{t}{\ln R_{ID}(t)}\right] \tag{8.15}$$

Substituting Equations (8.14) and (8.15) into Equation (8.10) yields

$$\begin{aligned} DC &= BC + SC_R\left[\left(-\left\{\frac{t}{\ln R_{ID}(t)}\right\}\right)\left(-\left\{\frac{\ln R_{SD}(t)}{t}\right\}\right)\right]^{\beta} \\ &= BC + SC_R\left[\frac{\ln R_{SD}(t)}{\ln R_{ID}(t)}\right]^{\beta} \end{aligned} \tag{8.16}$$

The above equation makes use of time dependent reliabilities of standard and improved designs instead of mean time between failures (*MTBF*s) as in the case of Equation (8.10).

8.2.6 Cost Estimation Model VI

This is concerned with estimating air filter energy cost [8]. The energy cost, *EC* in dollars, over the useful life of the filter is

$$EC = [(PC)(AQF)(FUL)(FR)K]/(10{,}000)(EMB) \tag{8.17}$$

where

EMB is the efficiency of motor and blower
FUL is the filter useful life expressed in hours
PC is the power cost expressed in \$/KWH
K is the constant with value 1.173
AQF is the air quantity to be filtered expressed in cfm
FR is the filter final resistance expressed in in. w.g.

8.2.7 Cost Estimation Model VII

This model does not directly estimate cost but it estimates required labor, hours/ton of product. This is a vital ingredient for estimating labor costs. The model was developed by Cevidalli and Zaidman [9]. The labor man-hours, *H*, per ton of product are expressed by

$$H = \frac{PK}{(1 + API)^n}\left[\frac{NFU}{(DC)^{\theta}}\right] \tag{8.18}$$

where

DC is the daily production capacity expressed in tons per day

θ is a constant:

$\theta = 0.86$ (for *DC* greater than 150)
$\theta = 0.76$ (for *DC* less than or equal to 150).

PK is the process factor (a measure of productivity):

$PK = 10$ (highly automated process)
$PK = 23$ (batch processes)
$PK = 17$ (continuous processes normally automated).

NFU is the quantity of functional units
n is the number of years since 1952
API is the annual productivity growth

8.2.8 Cost Estimation Model VIII

This model does not directly estimate cost but is developed [10] for determining the total number of employees, which in turn can be used to calculate labor cost. The total number of employees is given by

$$TNE = 7(INV)^{0.88} \tag{8.19}$$

where

TNE is the total number of employees
INV is the plant investment, in 1978, in millions of dollars

The above relationship is derived from data collected on 35 plants: ore plants, oil refineries, steel mills, etc. The employees include maintenance and support workers, plant workers, plant staff, and administration. Before using this relationship for a real life application, one should consult Reference [10].

8.2.9. Cost Estimation Model IX

This model was developed by A.V. Bridgwater [9] to estimate cost of utilities, *CU*, expressed in dollars per ton. The model correlates cost of utilities to the capital intensiveness of the plant and the quantity of functional units. The cost of utilities is given by

$$CU = B_1 M^{B_2} \left[\frac{CI}{PQ} \right]^{B_3} \tag{8.20}$$

where

PQ is the production quantity per year (tons per year)

CI is the fixed capital investment expressed in thousands of dollars
M is the quantity of functional units
B_i is the ith constant:

$i = 1$ ($B_1 = 0.748$)
$i = 2$ ($B_2 = 0.774$)
$i = 3$ ($B_3 = 0.751$).

8.2.10 Cost Estimation Model X

This model was developed by J.L. Viola [11] for making preliminary estimate of a plant's capital cost. The capital cost, CP, is correlated with production capacity expressed in pounds per year and plant complexity factor, PCF. For solid–fluid plants, the complexity factor is defined [2] by

$$PCF = [Q](CF)\alpha(1 - 0.6FR) \tag{8.21}$$

$$\alpha = 1.1067[(IOR)Q]^{0.0167} \tag{8.22}$$

where

α is the average throughput correction factor
CF is the correction factor associated with pressure level and special materials
Q is the quantity of major operating steps involved in the process
FR is the fraction of the Q steps handling solid–fluid mixtures (the value of FR should be equal to or less than 0.45)
IOR is the input-to-output weight ratio (i.e. raw material/product)

For stainless steel, the value of CF can be estimated from the following relationship:

$$CF = 0.6 + X(SF + 0.8) \tag{8.23}$$

$$X = 0.00045(PS) + [(2.5)10^{-6}](PS)^2 \tag{8.24}$$

where

PS is the pressure, in psi
SF is the stainless steel fraction

The battery-limits cost, CP, can be computed from the following equation [2]:

$$\log\left[\frac{CP}{A}\right] = \left[\frac{\log PCF}{A_1}\right]^{A_2} \tag{8.25}$$

$$A = A_3(PR)^{A_4}, \qquad \text{for } 2 \times 10^6 < PR < 50 \times 10^6 \tag{8.26}$$

where

PR is the annual production rate expressed in pounds
A_i is the ith constant:

$i = 1$ ($A_1 = 1.182$)
$i = 2$ ($A_2 = 2.1$)
$i = 3$ ($A_3 = 65$)
$i = 4$ ($A_4 = 0.6$).

8.2.11 Cost Estimation Model XI

Usually, the multiplexed transmission facilities' cost is non-linear. These facilities are established on a modular basis. More specifically, a fixed number of circuits (a module) share some equipment [12]. This model is concerned with approximating the cost, CTF, of providing facilities for k number of circuits. Such cost [12] is defined by

$$CTF = FC\left[\frac{(k-1)}{MS} + 1\right] + k(IC) \tag{8.27}$$

where

MS is the size of module
IC is the incremental cost per circuit
FC is the fixed cost for a module

8.2.12 Cost Estimation Model XII

This model is proposed in Reference [13] for estimating total problem-solving cost, $TPSC$, on a computer. The total cost [13] is given by

$$TPSC = CC + CCS + UTC + DC + AC \tag{8.28}$$

where

AC is the auxiliary charges cost (if any)
CCS is the cost associated with communications
UTC is the cost associated with user's time
CC is the cost associated with computing system
DC is the cost of delay

The computing system cost, CC, is given by

$$CC = (T) \cdot (CR) \tag{8.29}$$

where

T is the central processing unit time
CR is the charge rate for the central processing unit

The cost, DC, of delay is expressed by

$$DC = (ET)(DPC) \tag{8.30}$$

where

DPC is the delay penalty charge
ET is the elapsed time

The communications cost, CCS, is given by

$$CCS = (CT) \cdot (CCR) \tag{8.31}$$

where

CCR is the communication charge rate
CT is the connect time

8.2.13 Cost Estimation Model XIII

This model is concerned with estimating cost, $CERR$, of errors in a program with N errors. The cost of errors [14] is defined by

$$CERR = \sum_{i=1}^{N} (COE_i + CCE_i) \tag{8.32}$$

where

COE_i is the cost associated with occurrence of the error
CCE_i is the cost associated with correcting the error

The cost components associated with the cost of occurrence of the error are lost equipment time, wasted manpower, etc. The cost of correcting the error includes components such as supply cost, manpower cost, and equipment cost.

8.2.14 Cost Estimation Model XIV

This model is concerned with estimating software marketing cost. The marketing cost has two major components: the cost, CHO, associated with home office, and the cost, *CFS*, associated with field sales. The annual marketing cost, *AMC*, is expressed by

$$\begin{aligned} AMC &= CHO + CFS \\ &= CHO + COM + FS \end{aligned} \tag{8.33}$$

where

FS is the salaries, including overheads
COM is the commissions

The cost associated with field sales can be estimated from the following equation [15]:

$$CFS = (RC)(AS) + (1 + HR)[(QA)(SA)(QP) + (QP)(SS)] \tag{8.34}$$

where

HR is the overhead rate
AS is the annual sales
RC is the commission rate
QA is the number (quantity) of system analysts per salesman
SA is the system analyst's annual salary
QP is the number (quantity) of people involved in sales
SS is the salesperson's annual base salary

8.2.15 Cost Estimation Model XV

This model does not directly estimate cost but it estimates elapsed time for coding instructions. In turn, the elapsed time can be used to estimate costs. The elapsed time is defined [16] by

$$ET = \frac{IC}{(K)(NP)(QI)} \tag{8.35}$$

where

IC is the quantity of instructions to be coded
QI is the number of coded instructions per day per man
K is the constant (i.e., the number of work days per month). In this case the value of *K* is 22 days.
NP is the number of programmers
ET is the total time expressed in months

8.2.16 Cost Estimation Model XVI

Again, this model does not estimate cost directly but it estimates programmer man-months, which can be used to compute software related costs. The programmer man-months, PMM, are defined [17] by

$$PMM = \theta(DMC)^{\beta} \tag{8.36}$$

where

PMM is the programmer man-months
DMC is the thousands of lines of delivered machine code
θ is the constant
β is the constant with value 1.83

For assembly and high-order languages (HOL) the values of the θ are 59.2 and 28, respectively.

8.2.17 Cost Estimation Model XVII

This model is indirectly related to cost estimating and is concerned with estimating the time for constructing the proposed facility [18]. The time for constructing the proposed facility is given by

$$TPF = TEF\left[\frac{CPF}{CEF}\right]^{k} \tag{8.37}$$

where

k is the exponent. Its values range from 0.25 to 0.40
CEF is the existing facility's cost
CPF is the proposed similar facility's cost
TEF is the time taken for constructing the existing facility
TPF is the time for constructing the proposed similar facility

Example 8.1

A \$3 million project took 6 months to complete. A comparable new facility is proposed. The estimated cost of the proposed facility is \$5 million. If the value of k is 0.4, estimate the time for completing the new facility.

Substituting the given data into Equation (8.37) yields

$$TPF = 6\left[\frac{5}{3}\right]^{0.4}$$
$$= 7.3602 \text{ months}$$

Thus, the new facility will take 7.3602 months to complete.

8.2.18 Cost Estimation Model XVIII

This model is used to estimate cost of cooling towers [19], using operating data. The cost of a cooling tower, in dollars, can be estimated from the following relationships:

$$CCT = \frac{(HL)}{(Y)(TA) - 586 + (39.2)(CR)} \tag{8.38}$$

$$Y = \frac{279}{0.0335(85 - WBT)^{1.143} + 1} \tag{8.39}$$

$$CR = HWT - CWT \tag{8.40}$$

$$TA = CWT - WBT \tag{8.41}$$

where

CCT is the cost of a cooling tower
HL is the total heat load expressed in BTU per hour
CR is the cooling range
TA is the temperature approach
WBT is the wet bulb temperature expressed in degrees F
HWT is the hot water temperature expressed in degrees F
CWT is the cooled water temperature expressed in degrees F

Example 8.2

Estimate the cost of a cooling tower, if the following data are known:

$HL = 400$ million BTU per hour
$WBT = 75°\text{F}$
$CWT = 85°\text{F}$
$HWT = 115°\text{F}$

Using the above data we get

$$TA = 85 - 75 = 10°\text{F}$$

$$CR = 115 - 85 = 30°\text{F}$$

$$Y = \frac{279}{0.0335(85 - 75)^{1.143} + 1} = 190.3613$$

and

$$CCT = \frac{400{,}000{,}000}{(190.3613)(10) - 586 + (39.2)(30)} = \$160{,}409.82$$

Thus, the cost of the cooling tower in 1972 will be $160,409.82.

8.2.19 Cost Estimation Model XIX

This model is concerned with estimating the cost of heat exchangers [20]. The heat exchangers' cost estimation relationship is expressed by

$$HEC = \alpha(AA)^m \tag{8.42}$$

where

α and m are the constants
AA is the surface area of the heat exchanger in question expressed in ft^2
HEC is the procurement cost, free-on-board (f.o.b) factory

The values of α and m are given in Table 8.1 [20].

8.2.20 Cost Estimation Model XX

This model is proposed to estimate cost of storage tanks. The following relationship is developed [21] for calculating the cost:

$$STC = (MCF)(BC) \tag{8.43}$$

where

STC is the cost of a storage tank
BC is the base cost in carbon steel expressed in dollars
MCF is the material-of-construction factor

The base cost in carbon steel for field-erected and shop-fabricated tanks (cone roofs and flat bottoms), respectively, is given by

$$BC = \exp[F_1 - F_2 \ln VOL + F_3\{\ln VOL\}^2] \tag{8.44}$$

Table 8.1. Values for α and m.

No.	Heat exchanger type, construction material, and size range	m	α
1	Type: Spiral-plate Material: All type 304 stainless steel Size: 100–1500 (ft^2)	0.59	660
2	Type: Shell-and-tube Material: All type 304 stainless steel Size: 400–900 (ft^2)	0.67	235

for field-erected tanks, where

F_i is the ith constant:

$i = 1$ ($F_1 = 9.369$)
$i = 2$ ($F_2 = 0.1045$)
$i = 3$ ($F_3 = 0.045355$).

VOL is the tank volume in m^3: $80\ m^3 \leq VOL \leq 45{,}000\ m^3$

and

$$BC = \exp[S_1 + S_2 \ln VOL - S_3\{\ln VOL\}^2] \tag{8.45}$$

for shop-fabricated tanks, where

VOL is the tank volume in m^3: $5\ m^3 \leq VOL \leq 80\ m^3$
S_i is the ith constant:

$i = 1$ ($S_1 = 7.994$)
$i = 2$ ($S_2 = 0.6637$)
$i = 3$ ($S_3 = 0.063088$).

The values of MCF for construction materials such as aluminum, copper, stainless steel 316, stainless steel 304, stainless steel 347, titanium, monel, and nickel are 2.7, 2.3, 2.7, 2.4, 3, 11.0, 3.3, and 3.5, respectively.

8.2.21 Cost Estimation Model XXI

This model is concerned with estimating costs of centrifugal pumps. The cost of a pump [22] is defined by

$$CP = (PBC)(DTF)(MCF) \tag{8.46}$$

where

CP is the pump cost, including the base plate and driver coupling, but excluding the driver itself
PBC is the base cost for one-stage, 3,550 rpm, and vertically-split-case (VSC) pumps of cast iron. In this case $DTF = MCF = 1.0$.
DTF is the design-type cost factor
MCF is the construction material cost factor

The pump base cost can be estimated from the following relationship:

$$PBC = \exp[A_1 + A_2 \ln Z + A_3(\ln Z)^2] \tag{8.47}$$

Table 8.2. Values for a_1, a_2, and a_3.

Pump type	a_1	a_2	a_3
Multistage	2.08	−0.095	0.083
One-stage	0.461	−0.187	−0.025
Two-stage	0.745	−0.017	0.154

where

A_i is the ith constant:

$i = 1$ ($A_1 = 7.2234$)
$i = 2$ ($A_2 = 0.3451$)
$i = 3$ ($A_3 = 0.0519$).

Z is the size parameter

The size parameter Z is defined by

$$Z = (DC)(RH)^{1/2} \tag{8.48}$$

where

DC is the design capacity expressed in m^3/s.
RH is the required head expressed in J/kg.

The value of the design-type cost factor can be obtained from the following equation:

$$DTF = \exp[a_1 + a_2 \ln Z + a_3(\ln Z)^2] \tag{8.49}$$

For 3,550 rpm, horizontally-split-case (HSC) pumps, the values of a_1, a_2, and a_3 are tabulated in Table 8.2 [22].

The values of MCF for construction materials such as stainless steel 304, stainless steel 316, nickel, monel, titanium, ductile iron, bronze, and cast steel are 2.2, 3.5, 3.3, 9.7, 1.15, 1.9, and 1.35, respectively.

8.2.22 Cost Estimation Model XXII

This model is developed for estimating costs of pressure vessels [23]. The total cost, $TCPV$, of a pressure vessel is defined by

$$TCPV = (MCF)(BVC) + CPL \tag{8.50}$$

where

$TCPV$ is the total cost of pressure vessel expressed in dollars
MCF is the construction material cost factor
BVC is the base vessel cost in carbon steel, expressed in dollars
CPL is the platform and ladders cost expressed in dollars

For horizontal vessels, the BVC and CPL are expressed by Equations (8.51) and (8.52), respectively.

$$BVC = \exp[X_1 - X_2 \ln SW + X_3(\ln SW)^2] \qquad (8.51)$$

where

X_i is the ith constant:

$i = 1$ ($X_1 = 8.114$)
$i = 2$ ($X_2 = 0.16449$)
$i = 3$ ($X_3 = 0.04333$).

SW is the carbon steel shell weight in kg: 369 kg $\leq SW \leq$ 415,000 kg.

$$CPL = L_1 d^{L_2} \qquad (8.52)$$

where

L_i is the ith constant:

$i = 1$ ($L_1 = 1288.3$)
$i = 2$ ($L_2 = 0.20294$).

d is the platform's and ladder's inside diameter in m: 0.92 m $\leq d \leq$ 3.66 m.

Similarly, for vertical vessels, the BVC and CPL are expressed by Equations (8.53) and (8.54), respectively.

$$BVC = \exp[M_1 - M_2 \ln SW + M_3(\ln SW)^2] \qquad (8.53)$$

where

M_i is the ith constant:

$i = 1$ ($M_1 = 8.6$)
$i = 2$ ($M_2 = 0.21651$)
$i = 3$ ($M_3 = 0.04576$).

SW is the carbon steel shell weight in kg: 2210 kg $\leq SW \leq$ 103,000 kg.

$$CPL = m_1 d^{m_2}(TTL)^{m_3} \qquad (8.54)$$

where

m_i is the ith constant:

$i = 1$ ($m_1 = 1017$)
$i = 2$ ($m_2 = 0.73960$)
$i = 3$ ($m_3 = 0.70684$).

d is the platform's and ladder's inside diameter in m: $1.83 \text{ m} \leq d \leq 3.05$
TTL is the platform's and ladder's tangent-to-tangent length in m: $3.66 \text{ m} \leq TTL \leq 6.10$

The values of MCF for construction materials such as titanium, monel 400, stainless steel 304, inconel 600, nickel 200, stainless steel 316, and incoloy 825 are 7.7, 3.6, 1.7, 3.9, 5.4, 2.1 and 3.7, respectively.

8.2.23 Cost Estimation Model XXIII

This is an escalation-adjustment formula. The escalation may simply be described as the provision in estimated or actual costs for an augmentation in the costs of items such as materials, parts, and labor over the ones given in the initial contract [24]. Usually, escalation formulas to determine price adjustment are included in equipment manufacturers' bids. This formula is expressed [24] as follows:

$$ESC = (PR)(ORP)(PR_m)(RT_1) + (1 - PR_m)(ORP)(PR)(RT_2) \quad (8.55)$$

$$RT_1 = \frac{BLS_1}{BLS_2} \quad (8.56)$$

$$RT_2 = \frac{BLS_3}{BLS_4} \quad (8.57)$$

where

BLS_i is the ith U.S. Bureau of Labor Statistics (BLS) index:

$i = 1$ (material index at the time of shipment or end of fabrication)
$i = 2$ (material index at the time of placing order)
$i = 3$ (labor index at the time of shipment or end of fabrication)
$i = 4$ (labor index at the time of placing order).

ESC is escalation
ORP is the original price
PR_m is the percentage of the original price related to cost of materials
PR is the percentage of the original price which will be subject to escalation. More specifically, for this amount the equipment manufacturer has no control.

8.2.24 Cost Estimation Model XXIV

This model is used to estimate procurement cost of satellites in 1974 dollars. The procurement cost is defined by

$$SPC = [A_1(TW)^{-A_2}](TW) \tag{8.58}$$

where

SPC is the satellite procurement cost
TW is the total weight of the satellite
A_i is the ith constant:

$i = 1$ $(A_1 = 1970300)$
$i = 2$ $(A_2 = 0.592)$.

8.2.25. Cost Estimation Model XXV

This model is concerned with estimating total single-satellite system launch cost (circular orbits). The launch cost [25] is expressed by

$$TSC = B_1(SW)^{B_2}\left[\frac{AL}{B_3} + PSF + 1\right] \tag{8.59}$$

where

TSC is the total system cost for a single-satellite launch expressed in millions of dollars (assume 1974 dollars).
B_i is the ith constant:

$i = 1$ $(B_1 = 0.026)$
$i = 2$ $(B_2 = 2/3)$
$i = 3$ $(B_3 = 8{,}000)$.

AL is the orbit altitude or apogee expressed in statute miles
SW is the satellite weight expressed in pounds
PSF is a factor measuring the payload sophistication

8.3 Summary

This chapter has presented a large number of selective cost estimated models. Some of the models are only indirectly concerned with cost estimation. The models presented estimate the following:

i) Plant cost (Models I and X)
ii) Spares cost for the new aircraft system (Model II)
iii) Reliability acquisition cost (Model III)
iv) Cost per effective missile (Model IV)
v) Development cost (Model V)
vi) Air filter energy cost (Model VI)
vii) Labor man-hours (Model VII)
viii) Number of employees (Model VIII)
ix) Utility cost (Model IX)
x) Cost of providing facilities for *k* number of circuits (Model XI)
xi) Problem-solving cost (Model XII)
xii) Computer program errors' cost (Model XIII)
xiii) Software marketing cost (Model XIV)
xiv) Elapsed time for coding instructions (Model XV)
xv) Programmer man-months (Model XVI)
xvi) Time constructing the proposed facility (Model XVII)
xvii) Cooling tower cost (Model XVIII)
xviii) Heat exchanger cost (Model XIX)
xix) Storage tank cost (Model XX)
xx) Centrifugal pump cost (Model XXI)
xxi) Pressure vessel cost (Model XXII)
xxii) Escalation (Model XXIII)
xxiii) Satellite procurement cost (Model XXIV)
xxiv) Total single-satellite system launch cost (Model XXV)

8.4 Problems

1. What are the desirable features of a cost estimating model?
2. What is a cost estimating relationship (*CER*)?
3. Define the term "escalation".
4. Make comparisons of models I and X presented in this chapter.
5. Estimate the cost of a cooling tower using model XVIII and the following data:

$HL = 450$ million BTU per hour
$WST = 74°F$
$CWT = 84°F$
$HWT = 114°F$

References

1. J. Cran, Improved Factored Method Gives Better Preliminary Cost Estimates, Chemical Engineering, April 6, 1981, pp. 65–79.
2. T.J. Ward, Cost-Estimating Methods, in Design of Equipment, Vol. 1: Plant Design and Cost Estimating, Edited by J. Beckman, American Institute of Chemical Engineers, New York, 1986, pp. 12–21.
3. A.M. Tyszkiewicz, Comparative Cost Model Aircraft Investment Spares, Proceedings of the Annual Reliability and Maintainability Symposium, 1983, pp. 376–382.
4. E.S. Winlund, Cost-Effectiveness Analysis for Optimal Reliability and Maintainability, Proceedings of the Annual National Symposium on Reliability and Quality Control, 1965, pp. 107–114.
5. J.P. Kornfeld, Cost-Effectiveness as a Method of Evaluating Design Changes, IEEE Transactions on Engineering Management, March 1962, pp. 15–23.
6. R.R. Carhart, G.R. Herd, A Simple Cost Model for Optimizing Reliability, in Reliability of Miltary Electronic Equipment, a report by Advisory Group on Reliability of Electronic Equipment (AGREE), Office of the Assistant Secretary of Defense (Research and Engineering), Department of Defense, June 1957, pp. 64–74. Available from the Superintendent of Documents, U.S. Government Printing Office, Washington, D.C.
7. A.H. Hevesh, Cost of Reliability Improvement, Proceedings of the Annual Symposium on Reliability, 1969, pp. 54–61.
8. R.H. Avery, Life-Cycle Costing of High-Efficiency Air Filters, Plant Engineering, September 1977, pp. 80–83.
9. G. Cevidalli, B. Zaidman, Evaluate Research Projects Rapidly, Chemical Engineering, July 14, 1980, pp. 145–152.
10. R.B. Stein, Estimating Personnel for a New Plant, Chemical Engineering, Vol. 86, No. 20, September 24, 1979, p. 140.
11. J.L. Viola, Estimate Capital Costs Via a New, Shortcut Method, Chemical Engineering, April 6, 1981, pp. 80–86.
12. W.B. Elsner, Dimensioning Trunk Groups for Digital Networks, The Bell System Technical Journal, Vol. 59, September 1980, pp. 1123–1138.
13. D.N. Streeter, Cost-Benefit Evaluation of Scientific Computing Services, IBM Systems Journal, No. 3, 1972, pp. 219–233.
14. C. Sontz, Quality Assurance for the Data Processing Industry, Proceed-

ings of the Annual Reliability and Maintainability Symposium, 1973, pp. 136–148.
15. M. Phister, Data Processing Technology and Economics, The Santa Monica Publishing Company, Santa Monica, California, 1976.
16. L. Fried, Estimating the Cost of System Implementation—Part II, Data Processing Magazine, April 1969, pp. 24–28.
17. V. Schneider, Prediction of Software Effort and Project Duration—Four New Formulas, Sigplan Notices, Vol. 13, June 1978, pp. 49–59.
18. R.D. Stewart, R.M. Wyskida, Cost Estimator's Reference Manual, John Wiley & Sons, New York, 1987.
19. A. Zanker, Estimating Cooling Tower Costs from Operating Data, Chemical Engineering, June 12, 1972, pp. 118–120.
20. J.D. Kumana, Cost Update on Specialty Heat Exchangers, Chemical Engineering, June 25, 1984, pp. 169–172.
21. A.B. Corripio, K.S. Chrien, L.B. Evans, Estimate Costs of Heat Exchangers and Storage Tanks Via Correlations, Chemical Engineering, January 25, 1982, pp. 125–127.
22. A.B. Corripio, K.S. Chrien, L.B. Evans, Estimate Costs of Centrifugal Pumps and Electric Motors, Chemical Engineering, February 22, 1982, pp. 115–118.
23. A. Mulet, A.B. Corripio, L.B. Evans, Estimate Costs of Pressure Vessels Via Correlations, Chemical Engineering, October 5, 1981, pp. 145–150.
24. A. Pikulik, H.E. Diaz, Cost Estimating for Major Process Equipment, Chemical Engineering, October 10, 1977, pp. 106–122.
25. B.B. Hadfield, Satellite-Systems Cost Estimation, IEEE Transactions on Communications, Vol. 22, October 1974, pp. 1540–1547.

CHAPTER 9

Computer Hardware and Software Costing

9.1 Introduction

Today, the computer business is an important component of our national economy. For example, just over $57 billion was spent on computer systems in the United States government fiscal year 1980 [1]. The estimated computer hardware and software costs were in the order of $25.2 billion and $32 billion, respectively. More specifically, 56 percent of the total cost accounted for computer software and the rest for the computer hardware.

The software cost which accounts for a large proportion of the total computer system cost may be divided into two distinct categories: development (predelivery) cost and maintenance (post-delivery) cost. Several authors [2] have indicated different numbers for the proportions of these two costs. For example, in the case of maintenance cost the percentages were 50 [3], 75 [4], 60 [4], and 69 [2] of total software costs. The average of these four percentages is 63.5. Thus, one may say that roughly 63.5% and 36.5% of the total software cost account for maintenance and development, respectively. The above figures point out that a vast sum of money is being spent on computer systems and therefore computer software and hardware costing is of utmost importance.

There are various models and procedures available in the published literature to estimate cost of computer hardware and software. This chapter presents the topics of computer hardware and software costing. The emphasis of the chapter is on software costing.

9.2 Computer System Life Cycle Cost

As the cost of a computer system is not only confined to the procurement cost, one has to consider its life cycle cost in making related effective decisions. A major portion of the life cycle cost occurs during the use of a computer system. Thus, the life cycle cost of a computer system may be expressed as follows:

$$LCC_{cs} = PC + OC \tag{9.1}$$

where

LCC_{cs} is the life cycle cost of the computer system
PC is the procurement cost of the computer system
OC is the ownership cost of the computer system

Items such as system hardware, software license fees, installation, training, and documentation are generally included in the procurement cost. The computer ownership cost includes items such as preventive and corrective maintenance, supplies (e.g., power required to run the computer, paper for the printers, tapes and disks for the mass storage devices) and cost of the computer down time.

A more detailed simple model for estimating life cycle cost, LCC_c, of a computer may be expressed as follows:

$$LCC_c = PC_c + \sum_{j=1}^{k} (C_{cmj} f_j)/(1+i)^j \tag{9.2}$$

where

PC_c is the procurement cost of the computer
k is the life of the computer in years
i is the discount rate
f_j is the expected number of times the computer fails in year j
C_{cmj} is the single corrective maintenance call cost during year j

In this model, it is assumed that the only ownership cost involved is the cost of corrective maintenance. If expected number of failures occurring for each year is the same, the Equation (9.2) becomes

$$LCC_c = PC_c + f \sum_{j=1}^{k} C_{cmj}/(1+i)^j \tag{9.3}$$

where f is the expected number of computer failures per year

Example 9.1

Assume that the acquisition cost of a small computer is \$6,000. The computer has a 6-year life and its expected number of failures per million hours is 400. The only ownership cost involved is the cost of corrective maintenance. Each corrective maintenance call costs \$300. If the discount rate is 15%, estimate the life cycle cost of the computer.

The expected number of computer failures per year is given by

$$f = \frac{(400)(8{,}760)}{1{,}000{,}000} = 3.504 \text{ failures/year}$$

Substituting the above resulting value, as well as, the given data into Equation (9.3), we get

$$\begin{aligned} LCC_c &= 6{,}000 + (3.504)(300) \sum_{j=1}^{6} \frac{1}{(1.15)^j} \\ &= 6{,}000 + (1{,}051.20)(3.7845) \\ &= \$9{,}978.27 \end{aligned}$$

Thus, the estimated life cycle cost of the small computer is \$9,978.27.

9.3 Computer Hardware Costing

This section presents three cost estimation models taken from published literature.

9.3.1 Model I

This model is concerned with estimating computer hardware maintenance cost. The simple model [5] which will allow organizations to evaluate their maintenance costs and practices is presented below:

$$MMC = PMC + IC + CMC \tag{9.4}$$

where

MMC is the monthly maintenance cost
PMC is the cost of preventive maintenance
IC is the cost of inventory
CMC is the cost of corrective maintenance

The cost of preventive maintenance is given by

$$PMC = \frac{(CER)(EH)[SCRPT + CETT]}{SPMI} \tag{9.5}$$

where

CER is the customer engineer's rate per hour. This also includes spare parts usage rate

EH is the operating hours of the equipment expressed in hours per month
SCRPT is the scheduled preventive maintenance time of the customer engineer
CETT is the travel time of the customer engineer for performing preventive maintenance
SPMI is the scheduled preventive maintenance interval

The customer engineer's hourly rate may be estimated from the following expression:

$$CER = \frac{CEHP(1+\theta)}{FTM} + PC \tag{9.6}$$

where

PC is the hourly cost of parts
θ is the overhead rate
CEHP is the pay of the customer engineer per hour
FTM is the fraction of time customer engineer spends on maintenance. The fraction $(1 - FTM)$ of customer engineer's time is spent on items such as paper work, waiting, and training.

The cost of inventory is given by

$$IC = (ICR) \cdot (IV) \tag{9.7}$$

where

ICR is the inventory cost rate per month. This rate includes depreciation, handling costs, and interest charges for spares
IV is the maintenance spare parts inventory value. This value is the spare parts' original manufacturing cost.

Finally, the cost of corrective maintenance is defined by

$$CMC = \frac{(CER)(EH)[CMTT + MTTR]}{MTBF} \tag{9.8}$$

where

MTTR is the mean time to repair
MTBF is the mean time between failures. One should note here that the *MTBF* is to be given in terms of in-use time.
CMTT is the travel time of the customer engineer for corrective maintenance

9.3.2 Model II

This model is concerned with approximating the yearly labor expenditure for servicing a computer system. For computer servicing the largest expense is for labor. This expense depends on factors such as below:

i) Mean time to preventive maintenance
ii) Average cost of labor
iii) Mean time between failures
iv) Preventive maintenance interval
v) Travel times
vi) Mean time to repair

The annual labor expense is expressed as follows [6]:

$$ALC = LC(8{,}760)\left[\frac{(ATPM + TPM)}{T_{pm}} + \frac{(RCTT + MTTR)}{MTBF}\right] \quad (9.9)$$

where

$MTBF$ is the mean time between failures
$MTTR$ is the mean time to repair
T_{pm} is the average time between preventive maintenance
LC is the hourly labor cost
$RCTT$ is the travel time associated with a repair call
$ATPM$ is the average time taken to carry out preventive maintenance
TPM is the travel time associated with a preventive maintenance call

Example 9.2

Assume that we have the following data:

$$LC = \$70$$

$$MTBF = 2{,}000 \text{ hours}$$

$$MTTR = 3.5 \text{ hours}$$

$$T_{pm} = 3{,}000 \text{ hours}$$

$$ATPM = 5.25 \text{ hours}$$

$$TPM = 0.75 \text{ hours}$$

$$RCTT = 1.5 \text{ hours}$$

Estimate yearly labor expenditure for servicing a computer system using Equation (9.9). Substituting the above data into Equation (9.9), we get

$$ALC = (70)(8{,}760)\left[\frac{(5.25 + 0.75)}{3{,}000} + \frac{(1.5 + 3.5)}{2{,}000}\right]$$

$$= \$2{,}759.4$$

Thus, the approximate yearly labor expense involved in servicing a computer system is \$2,759.4.

9.3.3 Model III

This model is concerned with estimating the integrated circuit manufacturing cost. The integrated circuit manufacturing cost, $ICMC$, per chip is defined by [7]:

$$ICMC = TC + PC + PKC \tag{9.10}$$

where

TC is the test cost
PC is the cost of processing
PKC is the cost of packaging

The test cost is given by

$$TC = \frac{TFC}{TY} \tag{9.11}$$

where

TFC is the cost of testing, completed integrated circuit
TY is the test yield

The processing cost is defined by

$$PC = \frac{WPC}{(WY)(TY)(PY)(CH)} \tag{9.12}$$

where

WY is the wafer yield
TY is the test yield
PY is the packaging yield
WPC is the wafer processing cost
CH is the chips per wafer

Finally, the cost of packaging is expressed as follows:

$$PKC = \frac{(ICPC + ICCC)}{(PY + TY)} \tag{9.13}$$

where

ICPC is the integrated circuit package cost
ICCC is the integrated circuit connections cost

The symbols *PY* and *TY* were defined earlier.

9.4 Computer Software Costing

According to hardware and software cost trends shown in a 1972 United States Air Force study, the software cost, in 1955, accounted for less than 20% of the combined total of hardware and software costs and its projection for 1985 was approximately 80% of the total. In addition, in a July 1976 issue of "Newsweek", it was stated that the ratios of computer hardware costs to software costs were 4 to 1 and 1 to 4 in 1950's and 1976, respectively. This drastic change has led the way for careful attention in software cost estimation. Over the past several years many software cost estimating methods and models have surfaced. This section discusses several aspects of software costing.

9.4.1 Difficulties Faced in Estimating Software Costs

In estimating software costs, one faces many problems because software cost estimating is a complex process influenced by several factors. There are many factors for not getting good estimates of software costs. Some of those are given below [8]:

i) Shortage of sufficient historic data for making checks
ii) Poor comprehension of the software development process and maintenance
iii) Shortage of sufficient historic data for calibration purposes. The calibration may simply be described as a process through which a model is fitted to a specified cost estimating situation.
iv) Inhibition of project to project comparisons because of belief in project uniqueness
v) Poor comprehension of the management and technical constraints effects

9.4.2 Areas from Which Present Software Cost Estimating Methods Suffer

The existing software cost estimating methods suffer from areas such as follows:

i) The scope of the estimate lacks proper specification. More specifically what is included and what is not.
ii) Poorly defined objective of the estimate
iii) Poor definition and comprehension of the premises on which it has been established

9.4.3 Software Cost Estimating Steps

A software cost estimation activity is a miniproject and it should be treated so. More specifically, it has to be planned, reviewed, and followed up. There are a number of steps involved in software cost estimation. The basic ones are presented in Figure 9.1. This figure shows seven steps [9]: define objectives, make plans for necessary needed data and resources, identify precise software needs, carry out as much detail as possible, make use of several independent cost estimation methods and sources, compare and iterate results, and take follow up necessary measures. As many of these steps are self-explanatory only some of them are described below.

In establishing objectives, it will be helpful to pay attention to guidelines such as key objective to the requirements for decision making information, balance the objective of estimating accuracy for the various system elements of the cost estimates, review objectives as the process progresses, and make changes as necessary.

In the case of step 4, shown in Figure 9.1 one is concerned with working out as much detail as possible. Broadly speaking one may say that the more depth or detail to which one performs cost estimation, the more accurate will be the end estimates. The next step, i.e., step 5, calls for using many independent cost estimation methods and sources. The reason for using a combination of estimating methods is to avoid the weak points of a single technique and to take advantage of their combined strength. Some of the cost estimation methods are the expert opinions, top-down, analogy, bottom-up, and algorithmic models. All of these techniques are described in this chapter at a later stage. One specific important factor for making use of many independent methods is to compare their end estimates. The difference in those estimates provides opportunities to investigate the reasons for the difference. The remaining steps are described in detail in Reference [9].

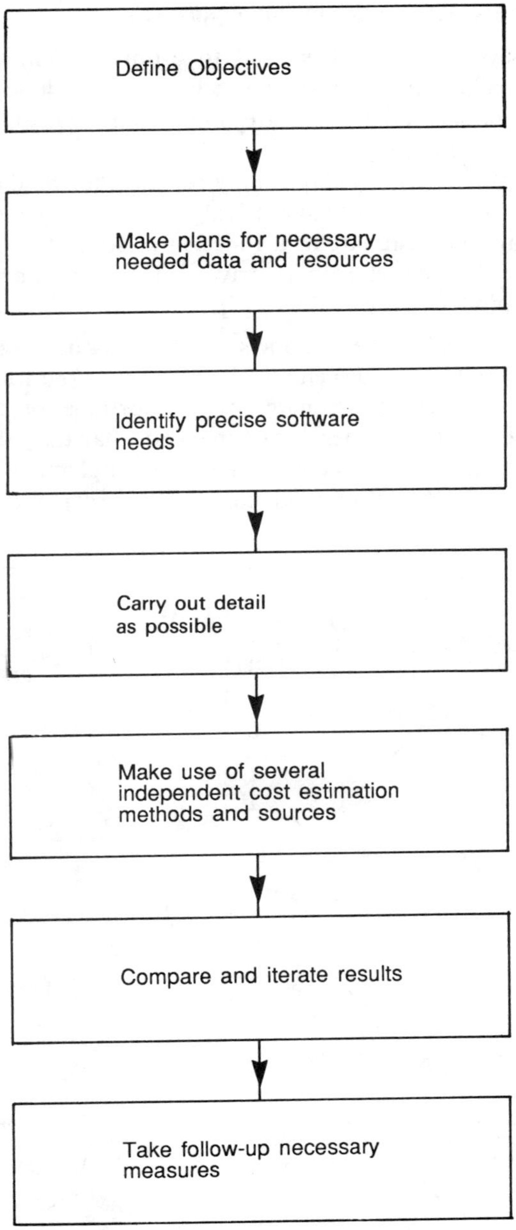

Figure 9.1. Basic steps for software cost estimating.

9.4.4 Software Cost Estimating Approaches

There are several approaches used to estimate computer software costs. They include the ones shown in Figure 9.2. These methods are algorithmic models, analogy, bottom-up estimating, expert opinions and top-down estimating.

The algorithmic models may simply be described as the ones which provide one or more mathematical algorithms for generating computer software cost estimate as a function of several variables. These variables are the important cost drivers. The five common types of algorithmic models are as follows [9]:

i) *Tabular models*: These models are composed of tables relating cost driver variables' values either to multipliers used for the purpose of adjusting the effort estimate, or to portions of the software development effort. In general, one may say that the tabular models are quite straightforward to comprehend and implement. Examples of these models are the Wolverton model [10], Aron model [11], and Boeing model [12].

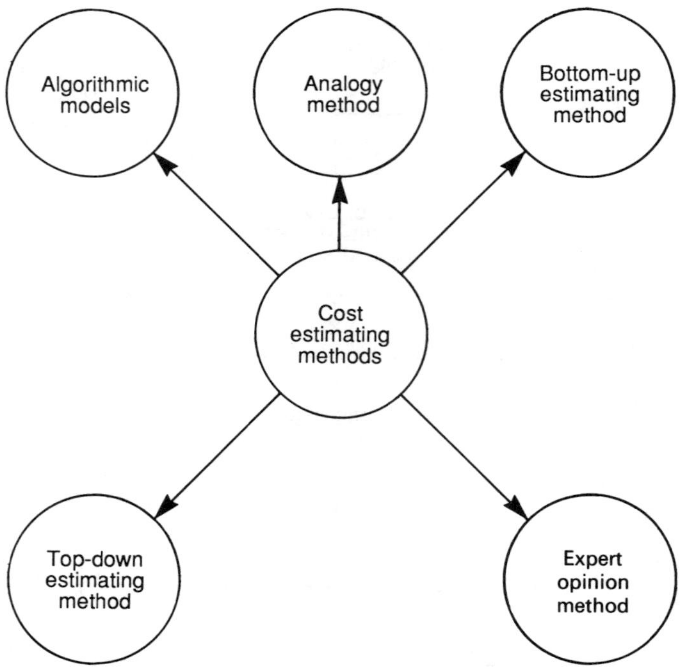

Figure 9.2. Software cost estimating methods.

ii) *Multiplicative models*: These cost estimating models take the following form:

$$EF = C_0 \prod_{i=1}^{m} C_i^{y_i} \tag{9.14}$$

where

m is the number of cost driver variables
EF is the effort
C_j is the coefficient selected to best fit the observational data, for $j = 0, 1, 2, 3, \ldots, m$
y_i is the cost driver variable i for $i = 1, 2, 3, \ldots, m$

For reasonably independent selected variables, this type of model seems to work fairly well. This type of model may be found in References [13, 14]. In order to obtain development cost one normally multiplies constant labor cost with the effort (EF) quantity.

iii) *Linear models*: The System Development Corporation software cost-estimation study [15] conducted in the mid-1960's used the linear model. The linear models take the following form:

$$EF = K_0 + \sum_{i=1}^{\theta} K_i r_i \tag{9.15}$$

where

θ is the number of cost driver variables
r_i is the cost driver variable i for $i = 1, 2, 3, \ldots, \theta$
K_j is the coefficient selected to best fit the observed data points; for $j = 0, 1, 2, \ldots, \theta$
EF is the effort

In concluding, it may be stated regarding the linear model that there are very many non-linear interactions in software development for it to perform very effectively.

iv) *Analytic models*: The existing models contain only a few variables. As a result they are not sensitive to a factor such as hardware constraints which often play an inportant role in software cost estimation. These cost-estimation models take the following general form:

$$RF = f(y_1, y_2, \ldots, y_m) \tag{9.16}$$

where

EF is the effort
f is a function (this function is neither multiplicative nor linear)

m is the number of cost driver variables
y_i is the cost driver variable i, for $i = 1, 2, 3, \ldots, m$

Halstead [16] and Putnam [17] models are two good examples.

v) *Composite models*: These models are relatively more difficult to learn and to use manually, as well as, require more data and effort. In order to determine software effort as a function of cost driver variables, the models in question incorporate an amalgamation of four types of functions: analytic, linear, tabular, and multiplicative. Two examples of composite models are the *TRW SCEP* [18] and *RCA PRICE-S* models [19].

The software cost estimation by analogy approach draws an analogy with the completed similar project. More specifically, it involves reasoning by analogy with accomplished similar software projects to relate their occurred costs to an estimate cost figure of the new project. This concept can be applied either at a subsystem level or at the total project level.

The bottom-up software cost estimating method involves decomposing of the total job into manageable components. Each component's cost is then estimated by an individual. Normally, the person assigned to develop the software component is also the one who estimates that component's cost. In order to obtain an estimate for the overall cost the costs of components are summed. A cost estimate obtained through the bottom-up method tends to be more stable. One of the weaknesses of this approach is that it may overlook system level costs. Examples of these costs are the costs associated with configuration management, integration, project management, and quality assurance. The expert opinion method is concerned with consulting experienced one or more experts to estimate cost of the proposed project. These experts make use of their expertise concerning the project under study to obtain its cost estimate. The Delphi technique is used in a number of studies to estimate software costs. It is an expert-consensus method. This method was developed in 1948 for predicting future occurrences by the Rand Corporation. Its name is derived from the location of the ancient Greek oracle.

The top-down software cost estimation method involves deriving a total cost estimate for the project, from the software product's global properties. The overall total cost estimate is then divided among the elements or components. The strength of the method are the efficiency and its system level focus. However, it is a less stable method. The method is complementary to the bottom-up approach and relatively requires less effort.

9.4.5 Software Life Cycle Cost Influencing Factors

There are a number of major factors which influence software life cycle costs. Their knowledge is useful for various purposes. For example, in making software management decisions, and in software cost estimation.

According to Reference [20], the major influencing factors may be categorized into five basic groups. These groups are shown in Figure 9.3. These are five different types of attributes: computer, personnel, size, project, and product. The computer attributes include turn-around time, speed and main storage constraints, and so on. The personnel attributes are probably the most important because they have a much larger impact on software costs than any other groups

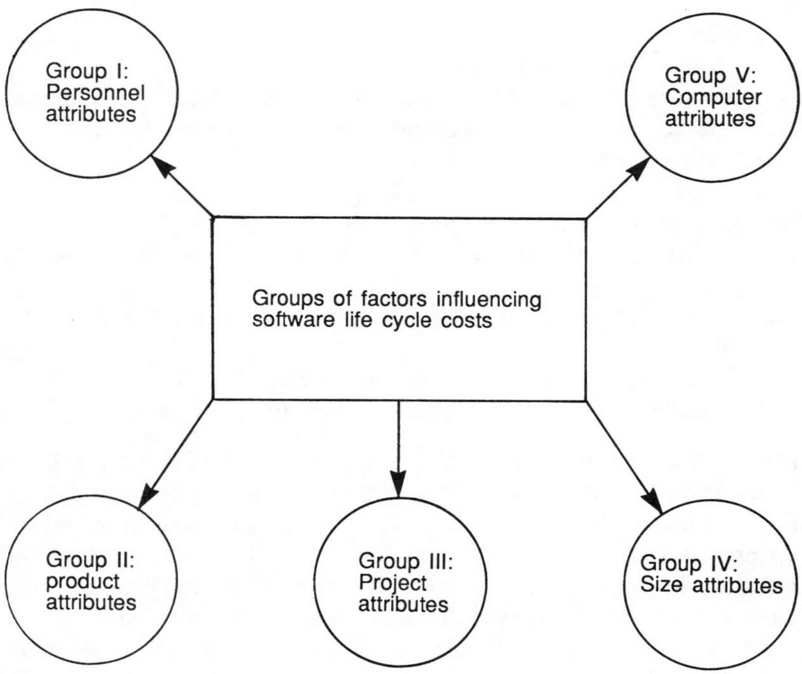

Figure 9.3. Basic categories of major factors influencing software life cycle costs.

of attributes. The personnel/team capability, teamwork, experience (programming language, virtual machine, applications), etc., belong to the personnel attributes group. The size attributes include a number of instructions, inputs, outputs, data elements, etc. The modern programming practices, use of software tools [21], and schedule constraints are the members of the project attributes group. The product attributes include the software product complexity, the choice of programming language, required software reliability, etc.

9.4.6 Characteristics of Software Cost-Estimation Models and Software Cost Historic Databases

There are a number of software cost-estimation models developed in the past. Their characteristics may vary quite considerably. In any case, there are several characteristics one should expect from a software cost estimating model to exhibit. Some of those are listed below [8]:

i) Wide applicability
ii) Clearly stated assumptions
iii) Includes estimates of how and when the resource will be required
iv) Checked against an acceptable number of historic projects
v) Well defined scope
vi) Simple to use
vii) Includes possibilities for sensitivity analysis
viii) Ability to make use of actual project data as these become available
ix) Generates a range of likely values for the quantity being determined
x) Favors inputs based on objective criteria
xi) Sensitive to all the appropriate parameters of a project

Each of the above characteristics is described in Reference [8].

The historic cost data are the backbone of software cost estimating. For example, a software development cost estimate will be based on historic data irrespective of the method used. Thus, the existence of historic databases of software costs is vital. In United States, there are a number of such databases. Some of them are listed in Table 9.1. A comprehensive list, of software cost historic databases, is given in Reference [8].

Table 9.1. Historic software cost databases.

Database number	Number of variables and observations, respectively	Developed by
1	4, 39	John Hopkins University, USA, 1975
2	8, 11	IBM, USA, 1975
3	?, 17	United States Air Force (USAF), Data Systems Design Center (DSDC)-PARMIS Data, USA, 1978
4	84, 18	Planning Research Corporation, USA, 1970
5	146, 22	Systems Development Corporation (SDC), USA, 1967
6	?, Approx. 400	Rome Air Development Center (RADC), USA

9.4.7 Software Life Cycle Cost Model

In this model, the life cycle cost is made up of seven individual cost components: analysis, documentation, operating and support, design, code and checkout, installation, test and integration. Thus, the life cycle cost is given by

$$LCC_s = C_A + C_D + C_{OS} + C_{DN} + C_{CC} + C_I + C_{TI} \qquad (9.17)$$

where

LCC_s is the software life cycle cost
C_A is the cost associated with software analysis
C_D is the cost associated with software documentation
C_{os} is the cost associated with software operating and support
C_{DN} is the cost associated with software design
C_{cc} is the cost associated with code and checkout
C_I is the cost associated with software installation
C_{TI} is the cost associated with software test and integration

The detailed software life cycle cost breakdowns structure is presented in Reference [22].

9.4.8 Software Cost Estimating Models

There are a number of models developed to estimate software costs. A large number of them are described in References [8, 9, 23]. Table 9.2 lists fourteen. Five software cost estimation models are presented below:

Model I

This software cost-estimation model is known as the Doty model, named after its developer [24, 25] the Doty Associates Inc. The model was developed in 1977 as a result of a study commissioned by Rome Air Development Center, Griffiss Air Force Base, New York. The Doty study is described in detail in References [24, 25]. This study makes use of several formulas. Some of these are as follows:

Formula I

This formula is concerned with estimating the total cost, TC_{SD}, in dollars, of software development (developer cost). This cost is composed of two cost components: primary cost and secondary cost. The primary cost is associated with labor whereas the secondary cost with other than labor, e.g., documentation, computer time. Thus, the total software development cost is given by

$$TC_{SD} = PC + SC \tag{9.18}$$

where

PC is the primary cost associated with software development, in dollars

SC is the secondary cost associated with software development, in dollars

The primary, cost associated with software development, is expressed by

$$PC = (MLR)(MP) \tag{9.19}$$

where

MLR is the mean labor rate of the software development manpower expressed in dollars per man-month. This includes costs of administration, general cost, overhead cost, and fees as appropriate.

MP is the manpower needed for the software development expressed in man-months. This includes for design, analysis, code, debug, test, and checkout.

Table 9.2. Software cost estimating models.

Number	Name	Developed in	Reference/References
1	The Doty model	1977 (by Doty Associates Inc.)	[24, 25]
2	The JPL model	1981 (by R.C. Tausworthe)	[26]
3	The Tecolote provisional model	1974 (by B.C. Frederic)	[27]
4	The Aerospace model	1975 (by Aerospace Corporation)	[23]
5	The Wolverton model	1974 (by R.W. Wolverton)	[10]
6	The Putnam model	1974 (by L.H. Putnam)	[28]
7	The IBM model	1975 (by J.C. Malone)	[8, 13]
8	The GRC model	1974 (by M.A. Tabach, M.C. Ditmore)	[29]
9	The Grumman model	1979 (by H.F. Dircks)	[30]
10	The SDC model	1967 (by System Development Corporation)	[31]
11	The NAVAIRDEV-CEN model	1971 (by Naval Air Development Center)	[32]
12	The Boeing model	1977 (by Boeing computer services)	[33]
13	The Bailey-Basili Meta model	1981 (by J.W. Bailey and V.R. Basili)	[34]
14	The ESD model	1975 (by Electronics Systems Division, Air Force Systems Command)	[35]

The secondary cost associated with software development can be estimated from the following equation:

$$\begin{aligned} SC &= \sum_{j=1}^{k} SRC_j \\ &= \beta(PC) \\ &= \beta(MLR)(MP) \end{aligned} \tag{9.20}$$

where

SRC_j is the cost associated with secondary resource j, expressed in dollars
k is the number of secondary resources
β is the ratio of secondary software development cost to primary software development cost

Substituting Equations (9.19) and (9.20) into Equation (9.18) yields

$$\begin{aligned} TC_{SD} &= (MLR)(MP) + \beta(MLR)(MP) \\ &= (MLR)(MP)[1 + \beta] \end{aligned} \tag{9.21}$$

Formula II

This formula is concerned with estimating the manpower needed, in man-months, for the software development. The total manpower needed is expressed by

$$MP = \theta S^{\alpha} \tag{9.22}$$

where

θ and α are constants
S is the estimated program size expressed in thousands of object words or source lines

It is to be noted from the above equation that it estimates manpower needs under the situation when little is known about the software development other than the program size. For different applications (object code and source code) the values of θ and α are given in Tables 9.3 and 9.4, respectively. The data given in both these tables are taken from References [24, 25]. As it can be seen from Tables 9.3 and 9.4, the value of α varies from 0.719 to 1.263. Similarly, the θ

Table 9.3. Values of θ and α for different applications (object code).

Application no.	Type of application	Estimated value of θ	Estimated value of α
1	*B*	2.895	0.784
2	*CC*	4.573	1.228
3	*U*	12.039	0.719
4	*ST*	4.495	1.068
5	*A*	4.790	0.991

B: Business, *CC*: Command and control, *U*: Utility, *ST*: Scientific, *A*: All

Table 9.4. Values of θ and α for different applications (source code).

Application no.	Type of application	Estimated value of θ	Estimated value of α
1	*B*	4.495	0.781
2	*CC*	4.089	1.263
3	*U*	10.078	0.811
4	*ST*	7.054	1.019
5	*A*	5.258	1.057

B: Business; *CC*: Command and control; *U*: Utility; *ST*: Scientific, *A*: All

varies from 2.895 to 12.039. The values of θ and α for the "all" application category, given in Tables 9.3 and 9.4, can be used for other applications than the ones presented in these two tables.

Formula III

This is also concerned with estimating the manpower needs, in man-months, for software development. In particular, this formula takes into consideration the impact of environmental factors. Thus, the formula is used to estimate manpower requirements when environmental factors are known. The formula is expressed below:

$$MP = \theta_1 S^{\alpha_1} \prod_{i=1}^{m} EF_i \tag{9.23}$$

where

MP is the manpower needed for the software development expressed in man-months
m is the number of environmental factors
EF_i is the constant reflecting the effect of ith environmental factor on the need for manpower
θ_1 and α_1 are constants
S is the estimated program size expressed in thousands of source lines

The values of θ_1, α_1, and EF_i are presented in References [24, 25]. In the Doty's study it was found that Equation (9.22) gives higher estimates for $S \geq 10{,}000$ and Equation (9.23) for $S < 10{,}000$.

Formula IV

This formula is concerned with estimating the size of the new software by analogy. In this case, it is assumed that the size and major functions of the previously developed program for the same type of application are known. Thus the size, S_N, of the new program in object words can be estimated from the following relationship:

$$S_N = \frac{(S_{PD}) \cdot (F_N^a)}{F_{PD}^a} \tag{9.24}$$

where

$a = 0.678$
F_N is the projected number of major functions to be carried out by the new program
S_{PD} is the size of the previously developed program in object words
F_{PD} is the number of major functions being performed by the previously developed program

Model II

This software cost estimation model is known as the Tecolote provisional model, named after the company where B.C. Frederic [27] developed it in 1974. The cost model is basically concerned with military software developments and makes assumptions that a usable operating system and the computer hardware already exist. In the development of this model seven basic factors were given consideration. These are computer costs, threat characteristics, delivered instructions, fast storage capacity, development costs, operational instructions, and development labor [36]. The software size is addressed in terms of operational and delivered instructions. The model is based on data obtained from five military programs: navy tactical software developments (three programs), military software developments of unspecified application (two programs).

From all these five programs, the data collected were total operating instructions, total computer hours, total delivered instructions, total labor, and total words fast storage. Although the model is simple to use, it is based on a small database. Therefore, its accuracy is indeed questionable. The model is described in detail in Reference [36].

Model III

This software cost estimation model is known as the Naval Air Development Center (NAVAIRDEVCEN) model. It is the result of a study [32] conducted by the Naval Air Development Center in March 1971. The model is comprised of several equations for the purpose of predicting total acquisition costs of research, development, test, evaluation, and production associated with future avionic computer systems. In order to convert the functional needs of the system into hardware and software needs and costs, the model is composed of 10 basic steps. The equation for estimating man-months of effort to develop the software is [23, 32]:

$$MP = 2.8A + 1.3B + 33C - 17D + 10E + F - 188 \qquad (9.25)$$

where

MP is the number of man-months required for the software development
F is the percentage of new instructions
D is the average programmer's experience with system, expressed in years
A is the machine language instructions in the delivered program expressed in thousands
E is the number of independent consoles in the delivered system
C is the number of document types generated
B is the contractors' traveled man-miles

The model is described in detail in References [23, 32].

Model IV

This software cost estimation model is known as the Putnam model named after its developer L.H. Putnam [17]. The model was developed from United States Army Software development data. For generating life cycle estimates of total manpower and time needed for arriving at important milestones of software projects, this model provides a methodology. However, the model does not take directly into consideration constraints inherent in the development process.

The Putnam SLIM model software package is commercially available [9]. Putnam makes use of the Rayleigh equation and therefore his model can be expressed as follows:

$$\begin{aligned} \dot{X} &= 2M\alpha t\, e^{-\alpha t^2} \\ &= \frac{2M}{2T_p^2}\, t\, e^{-\{t^2/2T_p^2\}} \\ &= \frac{M}{T_p^2}\, t\, e^{-\{t^2/2T_p^2\}} \end{aligned} \qquad (9.26)$$

where

t is the elapsed time in years since software project development started

M is the total number of man-years that will be expended to develop the system

α is the curve shape parameter. In physical terms it may be described as the "problem solving rate".

$\dot{X}$ is the manpower at time t expressed in man-years per year

T_p is time, in years, when $\dot{X}$ is a maximum

The definite integral of Equation (9.26) yields

$$X = M[1 - e^{-\alpha t^2}] \tag{9.27}$$

where

X is the cumulative man-years expended by the system at time t

The number of delivered source instructions, SI, is given by

$$SI = (TC)M^{1/3}T_p^{4/3} \tag{9.28}$$

where

TC is the "technology constant". For 1973-style and 1978-style developments, the typical values of TC are 4984 and 10,040, respectively. This model is described in detail in References [17, 36].

Model V

This software cost estimation model is based on V. Schneider's article [37] published in 1978. This paper presents a number of formulas for estimating software costs. Two of the formulas are presented below.

Formula I

This formula is concerned with predicting software project effort in situations when very little is known about the project except its expected delivered instructions. The programmer-months, MP, may be estimated from the following equation.

$$MP = (1.7)(F_a)(SCF)(DI) \tag{9.29}$$

where

F_a is the labor estimate adjustment factor given in decimal fraction [38]. Its values for poorly managed projects and under best circumstances are 2.9 and 0.435, respectively.

SCF is the software complexity factor. For trivial, moderately complex, and very complex softwares, the values of *SCF* are 1, 5, and 10, respectively.
DI is the delivered instructions expressed in thousands

Formula II

This is concerned with estimating software project duration. Schneider [37] presented the following equation for estimating the minimum project duration, $PD_{\min}$, under the assumption that the total hardware is available during the project life:

$$PD_{\min} = (MP)/SZ \tag{9.30}$$

where

SZ is the average staff size assigned to the software project under study
MP is programmer-months

An alternative formula for estimating software project duration, *PD*, in months is given in Reference [13]. In this case the project duration is defined by

$$PD = A_1(MP)^{A_2} \tag{9.31}$$

where

$A_1 = 2.47$ and $A_2 = 0.35$
MP is programmer-months

9.4.9 Single-Equation Models for Estimating Software Operational and Support Costs

According to Reference [39], one can expect at least 30% of system costs to be associated with operation and support of software. Annually, this figure will translate into billions of dollars. This section presents three single equation models for predicting software operational and support costs [39, 40].

Model I

This model was developed for the project $E-3A$ [40] and is known as the "Rule of Thumb" model. The equation for this model is based on data obtained from a number of United States Air Force projects. The man-years are given by

$$MY = (NI)(Y)/10{,}000 \tag{9.32}$$

where

MY is man-years
NI is the number of instructions
Y is years

Model II

This is known as the F-16 support model [40] and it consists of several parameters. However, it is reduced to the equation given below.

$$MP = A_3(NI) \tag{9.33}$$

where

MP is the man-months
A_3 is equal to 0.00712
NI is the number of instructions

Model III

This model was developed by General Research Corporation [40]. The equation for estimating cost, SMC, of correcting software in year OSY after delivery is

$$SMC = (1740)(NI)(Z)$$

where

$$Z = [(-0.00111)(OSY) + 0.01611]$$

The symbols OSY and NI denote the operation and support year considered and the number of instructions, respectively.

9.5 Summary

This chapter discussed computer hardware and software costing. The emphasis of the chapter is on software costing. A model for estimating life cycle cost of the computer system is presented.

The computer hardware costing covers three cost estimation models. These models are concerned with estimating computer hardware maintenance cost, annual labor expenditure for servicing a computer system, and integrated circuit manufacturing cost.

The topic of software costing is described in detail. Reasons for not getting good estimates of software costs are listed along with areas from which present software cost estimating methods suffer. Seven basic steps for estimating software costs are described. A number of software cost estimating methods are presented. These are algorithmic models, analogy, bottom-up estimating, expert opinions, and top-down estimating. The algorithmic models include five types: tabular, multiplicative, linear, analytic, and composite. Five basic categories of major factors influencing software life cycle costs are presented. These categories are as follows:

i) Personnel attributes
ii) Computer attributes
iii) Size attributes
iv) Project attributes
v) Product attributes

Six software cost historic databases are listed. A software life cycle cost model is presented along with five cost estimation models. Fourteen software cost estimation models are listed in Table 9.2. These include the Doty, Putnam, Tecolote, and NAVAIRDEVCEN models. Three single equation models for estimating software operational and support costs are given.

9.6 Problems

1. Write an essay on computer hardware and software costing.
2. A computer costs \$8,000 and has a 8-year expected life. Its expected number of failures per million hours is 350. The cost of corrective maintenance is the only ownership cost involved. If the discount rate is 14% and each corrective maintenance call costs \$350, compute the life cycle cost of the computer.
3. What are the factors for not getting good estimates of software costs?
4. Describe the basic steps involved in estimating software costs.
5. Discuss the following four software cost estimation models:

    i) The Tecolote model
    ii) The Doty model
    iii) The Wolverton model
    iv) The Aerospace model

6. What are the issues impacting software development costs?

7. What are the characteristics one should expect from a software cost estimating model to exhibit?
8. Describe the following two software cost estimating methods:

   i) Bottom-up estimating
   ii) Expert judgement

9. Comment on three-single-equation models for estimating software operational and support costs, presented in this chapter.

References

1. "Carter Faces Problems in Achieving His 1980 Budget Goals", Wall Street Journal, January 23, 1979, pp. 4–5.
2. W.D. Goeller, The Cost of Software Quality Assurance, Proceedings of the Annual Conference of the American Society for Quality Control, 1981, pp. 684–689.
3. D.S. Alberts, The Economics of Software Quality Assurance, Proceedings of the National Computer Conference, 1976, pp. 433–442.
4. B.W. Boehm, Software Engineering, Study Report, Prepared by TRW, Redondo Beach, California, 1976, pp. 19–59.
5. M. Phister, Analyzing Computer Technology Costs—Part II: Maintenance, Computer Design, October 1978, pp. 109–118.
6. D.P. Siewiorek, R.S. Swarz, The Theory and Practice of Reliable System Design, Digital Press, Bedford, Massachusetts, 1982.
7. M. Phister, Data Processing Technology and Economics, Santa Monica Publishing Company, Santa Monica, California, 1979.
8. M. Stanley, Software Cost Estimating, Royal Signals and Radar Establishment Memorandum No. 3472, Procurement Executive, Ministry of Defense, Malvern, Worcs., U.K., 1982.
9. B.W. Boehm, Software Engineering Economics, Prentice-Hall, Inc., Englewood Cliffs, New Jersey 07632, 1981.
10. R.W. Wolverton, The Cost of Developing Large-Scale Software, IEEE Transactions on Computers, June 1974, pp. 615–636.
11. J.D. Aron, Estimating Resources for Large Programming Systems, NATO Science Committee, Rome, Italy, October, 1969.
12. R.K.D. Black, R.P. Curnow, R. Katz, M.D. Gray, BCS Software Production Data, Report No. RADC-TR-77-116, Boeing Computer Services, Inc., March 1977, NTIS Report No. AD-A 039852, Available from National Technical Information Service (NTIS), Springfield, Virginia.
13. C.E. Walston, C.P. Felix, A Method of Programming Measurement, and Estimation, IBM Systems Journal, Vol. 16, 1977, pp. 54–73.
14. J.R. Herd, J.N. Postak, W.E. Russell, K.R. Stewart, Software Cost Estimation Study—Study Results, Report No. RADC-TR-77-220, Vol. I, Doty Associates, Inc., Rockville, MD, June 1977.

15. E.A. Nelson, Management Handbook for the Estimation of Computer Programming Costs, Systems Development Corporation, October 31, 1966. NTIS Report No. AD-A 648750. Available from National Technical Information Service (NTIS), Springfield, Virginia.
16. M.H. Halstead, Elements of Software Science, Elsevier, New York, 1977.
17. L.H. Putnam, A General Empirical Solution to the Macro Software Sizing and Estimating Problem, IEEE Transactions on Software Engineering, 1978, pp. 345–361.
18. B.W. Boehm, R.W. Wolverton, Software Cost Modeling: Some Lessons Learned, Journal of Systems and Software, Vol. 1, 1980, pp. 195–201.
19. F.R. Freiman, R.E. Park, PRICE Software Model—Version 3: An Overview, Proceedings of the IEEE Workshop on Quantitative Software Models, October 1979, pp. 32–41. IEEE Proceedings Catalog No. TH 0067-9.
20. B.W. Boehm, Software Life Cycle Factors, in Handbook of Software Engineering, edited by C.R. Vick and C.V. Ramamoorthy, Van Nostrand Reinhold Company, New York, 1984, pp. 494–518.
21. B.S. Dhillon, Reliability in Computer System Design, Ablex Publishing Corporation, Norwood, New Jersey, 1987.
22. M.E. Earles, Factors, Formulas, and Structures for Life Cycle Costing, Published by Eddins-Earles, 89 Lee Drive, Concord, Mass. 01742, 1981.
23. T.G. James, Software Cost Estimating Methodology, Proceedings of the IEEE 1977 National Aerospace and Electronics Conference (NAECON'77), 1977, pp. 22–28.
24. J.H. Herd, J.N. Postak, W.E. Russell, K.R. Stewart, Software Cost Estimation Study—Volume I, Report No. RADC-TR-77-220, 1977, Doty Associates, Inc., 416 Hungerford Drive, Rockville, MD 20850, U.S.A.
25. D.L. Doty, P.J. Nelson, K.R. Stewart, Software Cost Estimation Study—Volume II, Report No. RADC-TR-77-220, 1977. Doty Associates, Inc., 416 Hungerford Drive, Rockville, MD 20850, USA.
26. R.C. Tansworthe, Deep Space Network Software Cost Estimation Model, Report No. NASA-CR-164277, April 1981. Jet Propulsion Laboratories, Deep Space Network Data Systems Division, Pasadena, California.
27. B.C. Frederic, A Provisional Model for Estimating Computer Program Development Costs, Report No. TM-7/REV 1, 1974, Tecolote Research, Inc., Santa Barbara, California.
28. L. H. Putnam, A General Empirical Solution to the Macro Software Sizing and Estimating Problem, IEEE Transactions of Software Engineering, 1974, pp. 345–361.
29. M.A. Tabach, M.C. Ditmore, Estimation of Computer Requirements and Software Development Costs, Report No. RM-1873, March 1974. General Research Corporation, USA.
30. H.F. Dircks, SOFCOST Grumman's Software Cost Estimating Model, Proceedings of the IEEE 1981 National Aerospace and Electronics Conference (NAECON'77), 1981, pp. 100–109.

31. E.A. Nelson, Management Handbook for the Estimation of Computer Programming Costs, System Development Corporation, 2500 Colorado Avenue, Santa Monica, CA, 1967.
32. F. Buck, et al, A Cost by Function Model for Avionic Computer Systems, Report No. NADC-SD-7088, Vol. I, 1971. Developed by Naval Air Development Center, Warminster, PA 18974.
33. R.K.D. Black, R.P. Curnow, R. Katz, M.D. Gray, BCS Software Production, Report No. RADC-TR-77-116, 1977, Boeing Computer Services, Inc., Seattle, Washington 98124.
34. J.W. Bailey, V.R. Basili, A Meta-Model for Software Development Resource Expenditures, Proceedings of the 5th IEEE/ACM International Conference on Software Engineering, 1981, pp. 107–116.
35. Government/Industry Software Workshop Summary Notes, Electronic Systems Division, Air Force Systems Command, Hanscom Air Force Base, MA, 1975.
36. M.F. Finfer, R.K. Mish, Software Acquisition Management Guidebook: Cost Estimation and Measurement, Report No. ESD-TR-78-140, 1978, System Development Corporation, 2500 Colorado Avenue, Santa Monica, CA 90406.
37. V. Schneider, Prediction of Software Effort and Project Duration—Four New Formulas, Sigplan Notices, Vol. 13, 1978, pp. 49–59.
38. Embedded Computer Resources and DSARC Process—A Guidebook, Office of the Secretary of Defense, Room 2A-318, The Pentagon, Washington, D.C., 1977.
39. D.V. Ferens, R.L. Harris, Avionics Computer Software Operation and Support Cost Estimation, Proceedings of the IEEE 1979 National Aerospace and Electronics Conference (NAECON'79), 1979, pp. 296–300.
40. Advanced Avionics Systems for Multi-Mission Applications (AASMMA), Vol. II, Appendix G, Boeing Aerospace Company, Seattle, Washington, 1978.

CHAPTER 10

Life Cycle Costing in Aircraft Industry

10.1 Introduction

The increasing cost of aircraft systems over the past several years has led to growing concerns and appearance of various activities and concepts to decrease cost. These may have helped to slow the trend of increasing cost, nevertheless, the rising cost in the aircraft industry is still a factor to be overcome [1]. In other words, attainment of a suitable program for further decreasing the total life cycle cost of both the individual and total complex of systems still poses a challange to many professionals; engineers, systems managers, logisticians, designers, etc. Out of the many concepts and efforts introduced for curtailing the rising cost in the aircraft industry some have just been an effort for increasing awareness and others have been directed at very specific aspects of the problem (e.g., improving reliability).

Life cycle costing has had a profound effect in aircraft industry and is one of those industries where it is being practiced often. Over the past number of years, a large number of publications have appeared on life cycle costing in aircraft industry. This chapter is directly or indirectly concerned with life cycle costing and related areas.

10.2 Problems Challenging Management in Cost Reduction and Reasons for Budget Overshooting

There has been significant progress made in designing aircraft systems for reduced production and support cost and in other related areas. However, there still exist problems that have defied solutions and offer a management challenge. Their solutions will reward handsomely. Some of those problems are given below [1]:

i) Cuts and fluctuations in budgets and technical uncertainties
ii) The need to invest money at the initial stage in order to, hopefully, save a large amount later on
iii) Complexities of new systems
iv) Shortcoming in continuity in important management positions
v) The cosmic nature of life cycle costing
vi) Insufficient emphasis on the early stage of the development process
vii) Demotivating influence from contracting mechanisms, for example, exorbitant amount of paper work, inspections, reviews, and audits
viii) Intricacy in developing minimum but acceptable military needs for modern systems (if applicable)

The budget overshooting is a problem often faced both by the system manufacturer and the user and is related to cost. It was the problem which was also felt in the corridors of the United States Senate and the House of Representatives with respect to, in particular, cost of military programs. In 1969, the committees of the Senate and the House of Representatives Commissioned the Department of Defense for finding out reasons for budget overshooting and taking corrective measures. The Defense Department's investigation revealed the following four factors for the price overshooting [2]:

i) Planning variations (6%)
ii) Fluctuations in the economic situation (16%)
iii) Manufacturer (53%): modifications, rise in the cost of spares, training and documentation
iv) Modifications by the Department of Defense (25%)

10.3 Aircraft Life Cycle Cost

The life cycle cost of an aircraft can be broken down into four parts: Research, Development, Test, and Evaluation (RDT&E); Production; Initial Support; and Operations and Support. In the categories of both RDT&E and production, the overall cost can further be broken down into three parts, i.e., airframe, avionics, and engines. The costs associated with support equipment, data, spares, contractual training, and special equipment belong to the initial support category. More specifically, the initial support cost is divided into support equipment, data, spares, contractual training, and special support equipment. Finally, the operations and support cost can be divided into several parts, for example, base level maintenance, training, and operations personnel; engine repair (depot), component repair (depot). One should note here that the life cycle cost breakdown structure of an aircraft may vary from one type of aircraft to another, as well as, from one organization to another. The life cycle cost breakdown of a typical fighter aircraft is discussed below:

10.3.1 Life Cycle Cost of a Typical Fighter Aircraft

The life cycle cost of a typical fighter aircraft may be divided into following three parts [3];

i) Development cost
ii) Procurement cost
iii) Operations and support cost

According to Reference [3], usually for a typical fighter aircraft the development cost, procurement cost, and operations and support cost (15 year) account for approximately 10%, 35%, and 55% of the total cost, respectively. The development cost includes the following elements:

i) Test and evaluation cost
ii) Cost of data (e.g., stress reports and test reports)
iii) Design and development cost
iv) Flight test support cost (e.g., cost of spares, personnel, and ground support equipment)

The design, manufacturing, and testing activities account for approximately 90% of the development cost. The following factors drive the cost of the development activity:

i) Reliability and maintainability characteristics (for example, mean time between failures and mean time to repair)
ii) Physical characteristics such as weight and size
iii) Mission capabilities

The procurement cost is another major component of the life cycle cost of a fighter aircraft and is composed of the following:

i) The cost of initial support
ii) The cost of system project management
iii) The cost of test and evaluation

iv) The flyaway cost
v) The cost of facilities
vi) The cost of data

The flyaway cost include the cost of engine(s), airframe, and avionics. The examples of the data cost are the cost of training manuals and the cost of technical publications. The cost of spares, inventory entry and management, ground support equipment, and training and training equipment belongs to initial support cost. The initial spares cost, the flyaway cost, and the cost of ground support equipment account for approximately 95% of the procurement cost. The procurement cost drivers are as follows:

i) Training system requirements
ii) Reliability and maintainability characteristics
iii) Maintenance concept
iv) Mission capabilities
v) Number of bases

The components of the fighter's operations and support cost are as follows:

i) The cost of fuel
ii) The cost of personnel
iii) The cost of depot maintenance
iv) The cost of item management
v) The cost of modifications
vi) The cost of replenishing spares
vii) The cost of base maintenance material
viii) The cost of replacement training
ix) The cost of facilities

The cost components given below account for approximately 85% of the operations and support cost:

i) Fuel
ii) Depot maintenance
iii) Personnel requirements
iv) Base maintenance material
v) Replenishment spares

The factors which drive the fighter's operations and support costs are shown in Figure 10.1.

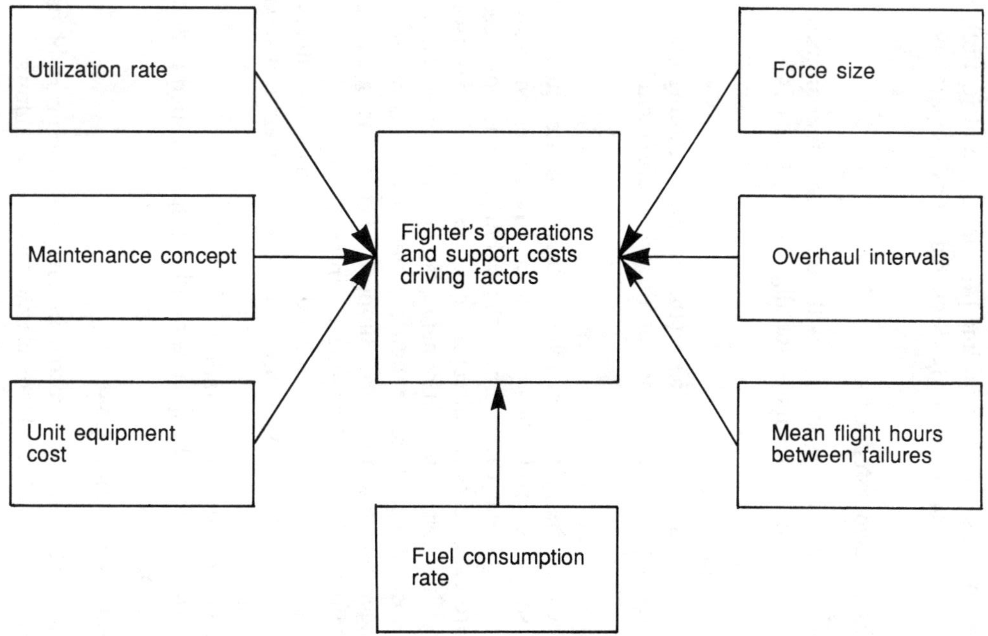

Figure 10.1. Fighter's operations and support costs driving factors.

10.4 Costs Involved in Air Transportation

This section discusses flight and load related costs. A cost structure which divides all costs into flight related and load related can be used by Airlines. At least one U.S. airline [4] makes use of it. Both flight related and load related costs are discussed below, separately.

10.4.1 Flight Associated Cost

This type of cost is associated with operating the aircraft. The following are the components of the flight related cost [4]:

i) The cost of fuel and oil
ii) The cost of insurance
iii) The cost of aircraft servicing (the aircraft handling costs such as refueling, cleaning, and landing fees are included in this category)
iv) The cost of direct maintenance
v) The cost of the flight crew (this includes expenses such as flight crew salary, payroll taxes, and the cost of retirement plans)
vi) The cost of the cabin crew (the cabin crew is composed of stewardesses, service director, etc.; associated expenses are salary, the cost of the retirement plan, payroll taxes, etc.)
vii) The cost of maintenance burden; this cost is associated with management and technical support of the line maintenance function for flight and ground equipment
viii) The cost of servicing administration (this includes supervising and administrative expenses associated with the aircraft servicing function)
ix) Depreciation and amortization
x) The cost of overhead expenses of all the flight related cash costs

10.4.2 Load Associated Cost

This is that cost which is incurred in carrying passengers or cargo. The components [4] of the load related cost are as follows:

i) The cost of reservations and direct sales
ii) The commissions paid on sales of cargo or passenger services to sales representatives or agents
iii) The costs related to passengers: liability insurance, food, supplies, and related administrative services

iv) The cost of advertising and publicity (this cost includes expenses incurred in all company advertising and publicity)
v) The cost of traffic servicing (this includes costs associated with cargo loaders, agents at the boarding gates and ticket counters, baggage handlers, etc.)
vi) The cost of supervising and administrative people involved in supporting the traffic servicing function
vii) The overhead expenses on all load associated cash costs listed above.

10.4.3 Flight and Load Related Costs of B-707 and B-747 Aeroplanes

The flight and load related costs may vary from one Airline to another and from one aeroplane to another aeroplane as below [4]:

The percentage breakdowns of flight and load related costs for a B-707 (Pan American World Airways) trip at a 55% seat factor/4,000 statute miles are as follows:

i) Flight associated cost (61%)
ii) Load associated cost (39%)

Table 10.1. Cost breakdowns for B-707 and B-747 aeroplanes.

No.	Cost component(s)	Percentage of total cost (load plus flight) for a B-707	Percentage of total cost (load plus flight) for a B-747
1	The cost of fuel and oil	10	9
2	The cost of traffic servicing	8	9
3	The cost of flight crew	13	7
4	The cost of maintenance and burden	13	11
5	The cost related to passengers (passenger variable)	9	10
6	All other costs	46	54

Further breakdowns of the above two costs (combined) for a B-707 are shown in Table 10.1. Similarly, the percentage breakdowns of flight and load related costs for a B-747 (Pan American World Airways) trip at a 55% seat factor/4,000 statute miles are given below.

i) Flight related cost (54%)
ii) Load related cost (46%)

Table 10.1 presents detailed breakdowns of the combined flight and load related cost associated with a B-747.

10.5 Cost Drivers

This section discusses cost drivers concerned with aircraft systems. In such systems, it is advisable to identify cost drivers so that the appropriate steps can be taken. This approach will be useful to reduce overall system cost. An example cited in Reference [5] states that 17% of purchased equipment parts contribute 90% of the cost for a specific type of aeroplane. This example clearly indicates the items for cost reduction emphasis.

According to historical data [5], in an aircraft system, the high cost areas are as shown in Figure 10.2. These are manpower (personnel), transportation/distribution, depot material, and bomb nav./propulsion/landing gear. Some of the high cost drivers are discussed in the following sections.

i) Facility locations
ii) Test concepts
iii) Unique skills
iv) Number of unique systems
v) Inspection concepts

10.5.1 Potential Cost Drivers in Two Typical Systems of a Combat Aircraft

The two typical systems presented here are fuel and hydraulic [6]. The potential cost drivers associated with a typical fuel system of a combat aircraft are listed below.

i) Valves
ii) Pumps
iii) Measurement
iv) Filters
v) Other components

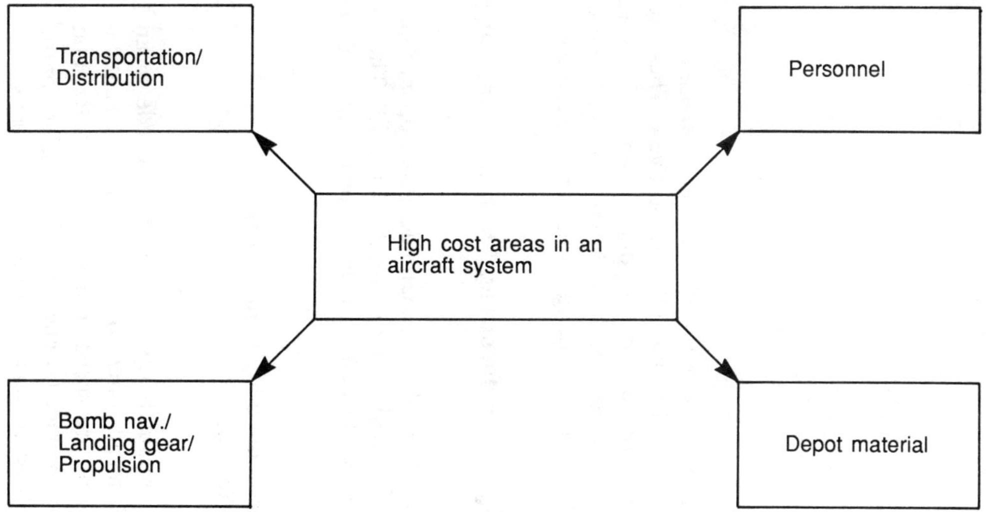

Figure 10.2. High cost areas in an aircraft system.

The expected percentage distribution of cost among above drivers is 33%, 27%, 17%, 8%, and 15%, respectively. In this case, over 80% of the cost is determined by valves, pumps, measurement, and filters. The valves and pumps account for 60% of the cost. Similarly the potential cost drivers associated with a typical hydraulic system are as follows:

i) Pumps
ii) Reservoirs
iii) Plumbing
iv) Filters
v) Valves
vi) Accumulators
viii) Other components

The expected percentage distribution of cost among these drivers is 26%, 20%, 12%, 12%, 9%, 7% and 14%, respectively. These figures indicate that over 85% of the cost is due to only six types of components: pumps, reservoirs, plumbing, filters, valves and accumulators. Almost half of the cost is due to pumps and reservoirs alone.

10.5.2 Typical Design, Manufacturing, Operations and Support Cost Drivers

This section discusses typical design, manufacturing, operations and support cost drivers associated with an aircraft system [3]. The typical design cost drivers can be divided into the following three different categories:

i) Reliability and maintainability requirements
ii) Performance requirements
iii) Specifications

The reliability and maintainability requirements include mean time between failures and mean time to repair. The elements of the performance requirement are speed, range, mission roles, and payloads. The specifications include fatigue life and corrosion control. The typical manufacturing cost drivers categories are as below:

i) Manufacturing process
ii) Structure (body and wing)
iii) Subsystems (flight control and landing gear)
iv) Material

The elements such as castings, sheet metal, forgings, and machined parts belong to the manufacturing process. The two major components of the structure are body and wing. The body includes cost drivers such as number of subsystems, wing attachment, and landing gear attachment. The wing includes number of hard points, complexity of control surfaces, and wet versus dry.

The subsystems have two major components: landing gear and flight control. The landing gear includes the number of wheels and brakes, and land versus carrier based. Similarly, the flight control includes number of redundancies, and mechanical versus fly-by-wire. The cost drivers belonging to the material category are steel, aluminum, composite, and titanium.

There are several operations and cost drivers of an aircraft system. The typical ones are shown in Figure 10.3. These are manpower, fuel, mean time between failures (*MTBF*), maintenance concept, mean time to repair (*MTTR*), and unit spares cost.

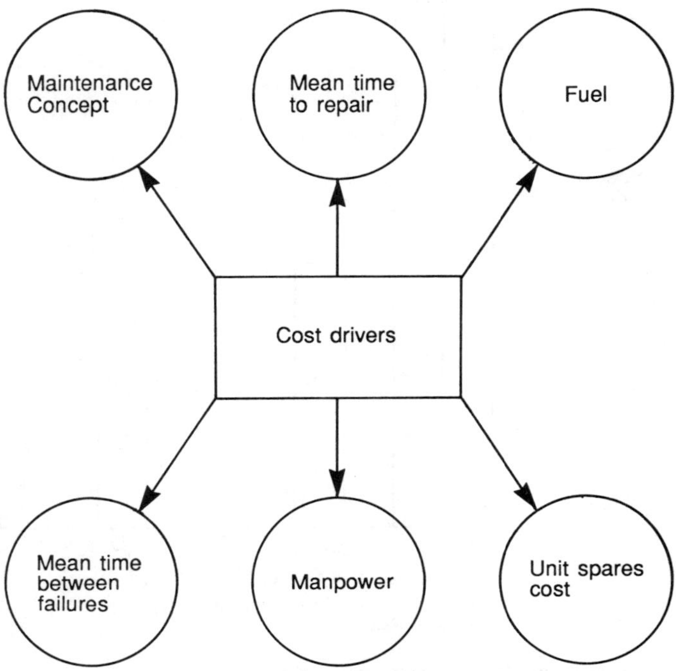

Figure 10.3. Typical operations and support cost drivers of an aircraft system.

10.5.3 Cost Drivers of the F-16 Multirole Fighter

The cost analyses revealed several cost drivers of the F-16 aircraft. According to these analyses, there were 50 cost drivers which accounted for approximately 82% of the total cost target [7]. Among the major divisions of the aircraft, the spread of the cost drivers was as shown in Figure 10.4. For example, this figure shows that the highest number (i.e., 22) of cost drivers were associated with airframe and the lowest (i.e., 2) with armament. The items listed below are the examples of cost drivers.

i) Fire and flight control computers
ii) Landing gear
iii) Inertial navigation set
iv) Nose radome

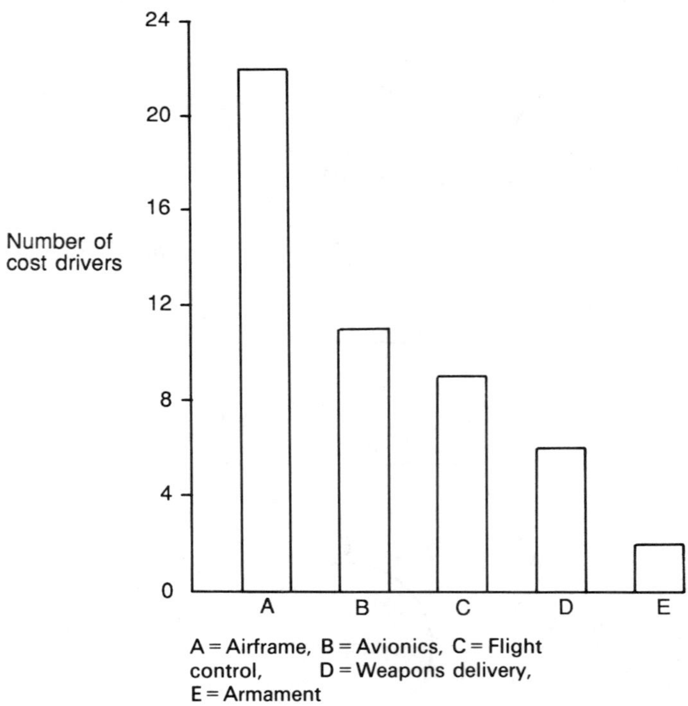

Figure 10.4. F-16 multirole fighter cost drivers.

10.5.4 Major Maintenance Cost Drivers for Helicopters

There are a number of maintenance cost drivers associated with helicopters. For example, maintenance cost drivers for military helicopters were as follows:

i) Power plants
ii) Inspections
iii) Rotor system (hub plus blades)
iv) Transmissions
v) Others

The breakdown percentages of direct maintenance cost (parts and labor) for each of the above categories were 27%, 9%, 29%, 12%, and 23%, respectively. The breakdown percentages within the rotor system were as follows:

i) Blades (80%)
ii) Hub (20%)

The major contributor to operation and support costs associated with rotor hub is the seal leak resulting in fluid and lubricant loss. Similarly, the rotor blade related contributors are foreign object damage and inability to repair damaged blades.

10.5.5 Major Aircraft Historical Problems and Airframe Maintenance Cost Components

The maintenance cost associated with an aircraft is directly or indirectly related to its historical problems. Therefore, identification of the top historical problems is useful. Some of the top ten aircraft historical problems are listed below:

i) Overheating of electronics
ii) Failure of hydraulic tubing and clamp
iii) Corrosion and fatigue of primary structure
iv) Inadequate fault isolation
v) Inaccuracy in fuel measuring system
vi) Teflon bearing failure

According to a 1973 analysis of airlines reports that the top ten airframe maintenance cost components are tires, nose landing gear wheel and tire, motor-driven hydraulic pump, coffee maker, constant speed drive, brakes, engine-driven hydraulic pump, flight control power control units, starter, and auxiliary power unit.

10.6 Models Employed in Estimating Life Cycle Cost for United States Air Force (USAF) Aircraft Systems

In the United States Air Force system acquisition process, the life cycle analysis has been gaining considerable attention over the past number of years. There are several models used in life cycle cost estimation for United States Air Force aircraft systems. The most widely used models are as follows [8]:

i) Logistics support cost model
ii) Programmed review of information for costing and evaluation (PRICE) model
iii) Depot Maintenance Cost Equations (Air Force Logistics Command)
iv) LCOM (Logistics Composite) model
v) Development and production costs of aircraft (DAPCA) model
vi) MOD-METRIC model
vii) Cost analysis cost estimating (CACE) and budgeting annual cost estimating (BACE) models
viii) Air Force manpower standards manual

Each of the above models is described below:

10.6.1 Logistics Support Cost Model

During aircraft design this is probably the most commonly used model [9] to generate operations and support cost estimates. The model was developed by the Air Force Logistics Command for the purpose of determining the difference between the support needs of some alternative aircraft designs. The maintenance processes, in this model, at the aircraft component level are represented by cost equations. There are ten equations that determine cost associated with various aspects of logistics support and many of them address to maintenance support and maintenance. At the level of subsystem and below, the model gives enormous cost visibility. The life cycle cost elements such as below are not addressed in this model [8]:

i) Health care
ii) Modifications
iii) Research and development
iv) Technical support
v) System investment
vi) Training ordnance
vii) Replacement ground support equipment
viii) Ordnance maintenance manpower
ix) Personnel support/permanent change of station

10.6.2 PRICE (Programmed Review of Information for Costing and Evaluation) Model

This model was developed by RCA for the purpose of calculating development and production costs of systems containing both mechanical and electronic parts. This is mainly an avionics cost model and addresses development costs related to the final configuration; the costs associated with research and early development are not taken into consideration [8]. For each application, the model generates necessary cost estimating relationships from its own data base. The use of the PRICE model has to be purchased from the RCA as she holds the proprietary right.

10.6.3 Depot Maintenance Cost Equations

These equations were developed by the Air Force Logistics Command for estimating depot maintenance cost based on performance and physical characteristics of the aircraft [10]. The examples of physical characteristics are speed, thrust, and weight. In order to develop these equations, the cost data utilized were obtained in part by apportioning engine overhaul and accessory repair costs to using aircraft system. There has been no validation attempt for these equations in Reference [10] and, therefore, the usefulness of their predictions is still a suspect. In pamphlet AFLC 173-4 [10] there are different equations for different types of aircraft: helicopter, fighter, trainer, and cargo.

10.6.4 The Logistics Composite Model

This is also known as the LCOM model and is a simulation model. The model represents the effects of support activities (base level) on flight operations of the aircraft. A number of support resources (such as spares, manpower, and support equipment) can be handled by the logistics composite model [11]. In determining maintenance manpower needs for new aircrafts, the Air Force has made use of it quite frequently. One should note here that the model in question requires a large quantity of input data. The detailed description of the logistics composite model is given in References [8, 11].

10.6.5 Development and Production Costs of Aircraft (DAPCA) Model

Just as for the PRICE model, the concern of the DAPCA model is research and development, and system investment. Under the acronym DAPCA, the Rand Corporation of Santa Monica, California has developed a series of models [8]. For example, the report on DAPCA-III [12] was published in March 1976. The basis for DAPCA-III is regression equations for turbine engines [13] and airframes [14] developed in 1974 and 1976, respectively. The DAPCA model does not estimate avionics systems costs and, in addition, has little sensitivity to factors involved in detailed design changes [8]. The major input variables for DAPCA model are speed, engine thrust, and weight. Thus, the model estimates procurement cost as a function of the following factors:

i) Speed
ii) Weight
iii) Mission type
iv) Thrust

The quality control methodologies used by this model are different for cargo aircraft and other types of aircrafts considered.

10.6.6 The MOD-METRIC Model

This is another type of cost estimation model used by the Air Force. The MOD-METRIC: a multi-item, multi-echelon, multi-indenture inventory model is frequently used to conduct detailed analyses of recoverable spares requirements (inventory) for a new aircraft systems [15]. The model is an extended version of METRIC (multi-Echelon Technique for Recoverable Item Control) model developed by Rand Corporation [16]. The inventory computations of this model (i.e.,

MOD-METRIC) relate to the following components of the life cycle cost:

i) Spare engines
ii) War readiness material
iii) Initial spares and repair parts

Even though the model in question was originally developed for use in logistics, life cycle costing is another area for which it has been also adopted. The detailed description of the model can be found in Reference [8].

10.6.7 Cost Analysis Cost Estimating (CACE) and Budgeting Annual Cost Estimating (BACE) Models

Both these models (CACE and BACE) used by the United States Air Force (USAF) are concerned with an aircraft squadron's operating and support costs [17]. Both models estimate costs for a typical aircraft squadron by making use of the four prime inputs given below:

i) Number of flying hours, H_F, per aircraft per year
ii) Manpower needs per aircraft squadron
iii) Unit equipment, E_u, aircraft per squadron
iv) Numerous cost factors (in order to obtain cost estimates these factors are multiplied by the number of men, flying hours (H_F), or unit equipment (E_u))

From the usage point of view the BACE and CACE models have the following purposes, respectively:

i) The BACE model is utilized for producing estimates for excercises performed as part of the Air Force Planning, Programming, and Budgeting System known as the PPB system
ii) The CACE model is utilized in life cycle costing, research or cost analyses, etc.

Both the models under consideration do not address the following life cycle cost elements [8]:

i) Initial investment cost
ii) Depot supply cost
iii) Research and development cost
iv) Second destination transportation cost

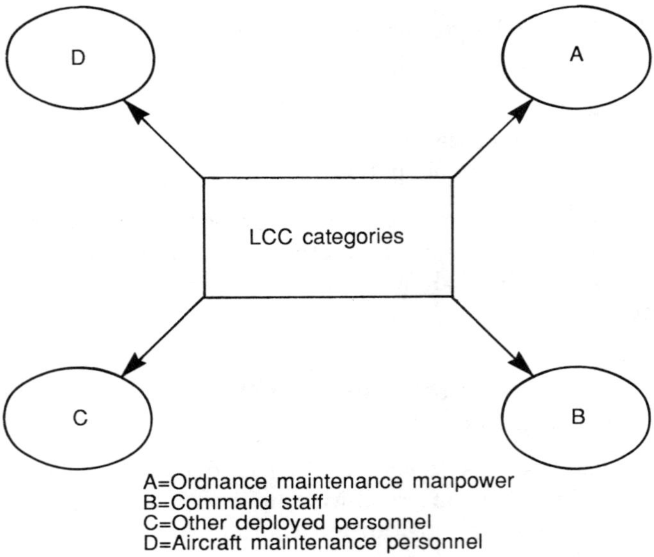

Figure 10.5. Life Cycle cost (LCC) categories addressed in the Air Force Manpower standards manual.

10.6.8 Air Force Manpower Standards Manual

This document [18] is composed of standards approved by the Department of United States Air Force (USAF) for the purpose of establishing detailed manpower authorizations. The Directorate of Manpower and Organization at major command and USAF headquarters makes use of these standards for developing manpower authorizations for all appropriate work centers. These standards address the life cycle cost elements [8] shown in Figure 10.5.

10.7 Cost Considerations in Aircraft Turbine Engines

The aircraft turbine engine is an important subsystem in the life cycle cost of an aeroplane and, therefore, is a particularly promising topic for study. There are various factors for their (defense aircraft engines) study. Some of those are listed below [18]:

i) A substantial inventory and budgetary expense
ii) Their (engines) extreme importance in weapon-system applications

iii) Existence of a sizable data base for performing analysis
iv) The placing subsystems in weapon-system development

The engine (including fuel) contribution to the complete combat aircraft life cycle cost might be considered as approximately 30% [19]. This figure is subject to the following assumptions:

i) The aircraft considered include development of a new engine and avionics system, as well as, airframe
ii) The aircraft is a single seat with twin engines
iii) A modest purchase of approximately 300 aircraft and 15 year peace time operation

The airframe and avionics account for 40% and 30%, respectively, of the combat aircraft life cycle cost. In addition, the approximate distribution of the engine life cycle cost is as follows:

i) Development (40%)
ii) Production (23%)
iii) Support excluding fuel (37%)

10.7.1 Basic Causes of Engine Failures

During the operation and support phase, hardware failures are a significant cost driver. Thus, the understanding of these failures is useful in life cycle costing. There are several causes of turbine engine failures. Some of these are as follows [21]:

i) Underdesign
ii) Thermal and mechanical fatigue (low cycle)
iii) Inadequate development
iv) Over temperature
v) Foreign object damage
vi) Creep
vii) Thermal and mechanical fatigue (high cycle)
viii) Corrosion
ix) Manufacturing defects
x) Stress rupture

10.8 Cost Estimation Models Used in Aircraft Industry

This section presents a number of cost models developed for use in the aircraft industry. These models were selected from various sources.

10.8.1 Life Cycle Cost Model for an Aircraft Turbine Engine

This model is concerned with estimating the life cycle cost of aircraft turbine engines. The engine life cycle cost is expressed as follows:

$$LCC_{TE} = C_1 + C_2 + C_3 + C_4 + C_5 \qquad (10.1)$$

where

LCC_{TE} is the life cycle cost of the aircraft turbine engine
C_1 is the base maintenance cost
C_2 is the cost of development
C_3 is the part improvement cost
C_4 is the depot maintenance cost
C_5 is the production quantity cost

Equations for estimating C_1, C_2, C_3, C_4, and C_5 are given in Reference [22].

10.8.2 Life Cycle Cost Model for a Tactical Aircraft

This model is concerned with estimating life cycle cost of tactical aircraft having speeds of $1.2 \leq M \leq 2.0$. In this model, the life cycle cost is composed of three major components: research, development, test, and evaluation; investment cost; and operating and support cost. Thus, the life cycle cost of the aircraft in question is expressed [23] by:

$$LCC_{TA} = C_{RD} + C_I + C_{OS} \qquad (10.2)$$

where

LCC_{TA} is the tactical aircraft life cycle cost in 1977 dollars
C_{RD} is the cost of research, development, test, and evaluation ($RDT\&E$)
C_I is the investment cost
C_{OS} is the operating and support cost

The following relationship is used for estimating research, development, test, and evaluation cost:

$$C_{RD} = aC_F^b \qquad (10.3)$$

where

$a = 0.376$
$b = 1.411$
C_F is the flyaway cost for unit 750

The investment cost, C_I, is given by

$$C_I = C_{IT} + C_{IS} + C_{SE} + F_{PQ}C_F\alpha_Q + C_{ITD} \qquad (10.4)$$

where

C_I is the investment cost
C_{IT} is the cost of initial training
C_{IS} is the cost of initial spares
C_{SE} is the cost of support equipment
C_F is the mean flyaway unit cost at quantity 750
F_{PQ} is the production quantity factor
C_{ITD} is the cost of initial technical data
α_Q is the quantity to be manufactured

Finally, the operating and support cost, C_{OS}, is expressed as

$$C_{OS} = \sum_{i=1}^{3} C_i \qquad (10.5)$$

where

C_1 is the cost of maintenance
C_2 is the cost of oil, lubricants, and petroleum
C_3 is the other support cost

Equations for estimating C_F, C_{IT}, C_{IS}, C_{SE}, C_{ITD}, C_1, C_2 and C_3 are presented in Reference [23].

10.8.3 Equations for Determining Aircraft Depot Maintenance Costs

This section presents a set of parametric equations for estimating aircraft depot maintenance costs. These equations were developed in Reference [24] for the purpose of application in determining the depot maintenance cost of new United States Air Force (USAF) aircrafts. The explanatory variables in equations presented in this section had values in the data base within ranges: 367–7475 lbs (engine dry weight), 4067–320,085 lbs (airframe empty weight), and 3400–65,840 psf (engine pressure term). Before using the equations given below in real life applications, the reader should consult Reference [24].

Equation I

This equation is concerned with estimating yearly engine component and accessory repair cost. The cost is expressed below:

$$C_{ECAR} = \alpha_e(P_E)^{\beta_e}(W_{ED})^{\theta_e} \tag{10.6}$$

where

C_{ECAR} is the engine component and accessory repair cost per engine per year
P_E is the engine pressure term (psf)
W_{ED} is the engine dry weight (lbs), and
$\alpha_e = 0.0265$
$\beta_e = 0.778$
$\theta_e = 0.677$

Equation II

This equation is concerned with estimating average engine cost per overhaul. The equation is given below:

$$C_{ACO} = \alpha_0 P_E^{\beta_0} W_{ED}^{\theta_0} \tag{10.7}$$

where

C_{ACO} is the average engine cost per overhaul
$\alpha_0 = 0.598$
$\beta_0 = 0.793$
$\theta_0 = 0.390$

The symbols P_E and W_{ED} are the same as defined for Equation (10.6).

Equation III

This equation is concerned with estimating yearly airframe rework (i.e., work associated with whole aircraft rather than individual components, but excluding engines) cost per aircraft and is expressed as follows:

$$C_{AFR} = \alpha_a W_E^{\beta_a} M_{pd}^{\theta_a} \tag{10.8}$$

where

C_{AFR} is yearly airframe rework cost per aircraft
W_E is the aircraft empty weight in lbs
M_{pd} is the programmed depot maintenance ($1 =$ no M_{pd}; $2 =$ has M_{pd})
$\alpha_a = 183$
$\beta_a = 0.344$
$\theta_a = 3.22$

Equation IV

This equation is concerned with estimating annual cost to repair an engine. Thus, the yearly cost to repair (per engine) is defined by

$$C_{YER} = \alpha_y P_E^{\beta_y} W_{ED}^{\theta_y} \tag{10.9}$$

where

C_{YER} is the yearly cost to repair (per engine)
$\alpha_y = 2.72 \times 10^{-8}$
$\beta_y = 1.49$
$\theta_y = 1.24$

The symbols P_E and W_{ED} are the same as defined for Equation (10.6).

Equation V

This equation is concerned with estimating annual airframe component repair cost per aircraft and is expressed as follows:

$$C_{CAR} = \alpha_c W_E^{\beta_c} \tag{10.10}$$

where

C_{CAR} is the yearly airframe component repair cost per aircraft
$\alpha_c = 0.788$
$\beta_c = 0.967$

The symbol W_E is the same as defined for Equation (10.8).

Equation VI

As the mean time between engine overhauls plays an important role in cost estimation, this equation is concerned with estimating aircraft engine mean time between overhauls. The mean time between overhauls is given by

$$MTBO_E = \alpha_m P_E^{-\beta_m} D_M^{\theta_m} \tag{10.11}$$

where

$MTBO_E$ is the aircraft engine average time between overhauls, in hours
D_M is the mission designator (1 = bomber/cargo; 2 = fighter/attack)
$\alpha_m = 957{,}000$
$\beta_m = 0.601$
$\theta_m = 1.23$

The symbol P_E was defined earlier.

10.9 Summary

This chapter has presented several aspects of life cycle costing in the aircraft industry. The problems challenging management in cost reduction and reasons for budget overshooting are presented. The life cycle cost of a typical fighter aircraft is discussed. The fighter's operations and support cost driving factors are shown. The flight and load related costs involved in air transportation are discussed in detail. In particular, the flight and load related costs of the B-707 and B-747 aircraft are given. Typical design, manufacturing, operations and support cost drivers are described. The cost drivers associated with the F-16 multirole fighter are given. Eight models employed in estimating life cycle cost for United States Air Force (USAF) aircraft systems are discussed. These are logistics support model, PRICE model, depot maintenance cost equations, LCOM model, DAPCA model, MOD-METRIC model, CACE and BACE models, and the Air Force Manpower Standards Manual. The cost considerations in aircraft turbine engines are discussed. Turbine engine failure basic causes are listed. A number of cost estimation models are presented. These models include life cycle cost models for turbine engine and tactical aircraft, and parametric equations for estimating aircraft depot maintenance costs.

10.10 Problems

1. Discuss the following:

   i) USAF Manpower Standards Manual
   ii) DAPCA model
   iii) CACE and BACE models

2. Who developed the PRICE model? What is the area of its applications?
3. Make comparisons of logistics support cost and DAPCA models.
4. List top ten airframe maintenance cost components.
5. List cost drivers associated with typical aircraft hydraulic and fuel systems.
6. Discuss high cost areas in an aircraft system.

References

1. W.E. Lamar, A Review and Assessment of System Cost Reduction Activities, in Design to Cost and Life Cycle Cost, North Atlantic Treaty Organization (NATO) Advisory Group for Aerospace Research and Development (AGARD) Conference Proceedings No. 289, May 1980, pp. 4.1–4.38. Available from the National Technical Information Service (NTIS), Springfield, Virginia 22161, USA.
2. R. Tassinari, Organizing a Design-To-Cost Program, in Design to Cost and Life Cycle Cost, North Atlantic Treaty Organization (NATO) Advisory Group for Aerospace Research and Development (AGARD) Conference Proceedings No. 289, May 1980, pp. 13.1–13.14. Available from the National Technical Information Service (NTIS), Springfield, Virginia, 22161, USA.
3. E. Huie, H.F. Harris, Balanced Design–Minimum Cost Solution, in Design to Cost and Life Cycle Cost, North Atlantic Treaty Organization (NATO) Advisory Group for Aerospace Research and Development (AGARD) Conference Proceedings No. 289, May 1980, pp. 18.1–18.8. Available from the National Technical Information Service (NTIS), Springfield, Virginia 22161, USA.
4. W.C. Messecar, Role of Life Cycle Costing in Fleet Planning Decisions, Proceedings of the 10th Reliability and Maintainability Conference, Anaheim, California, 1971, pp. 236–240.
5. F.T. Carlson, Design to Life Cycle Cost Research, in Design to Cost and Life Cycle Cost, North Atlantic Treaty Organization (NATO) Advisory Group for Aerospace Research and Development (AGARD) Conference Proceedings No. 289, May 1980, pp. 5.1–5.15. Available from the National Technical Information Service (NTIS), Springfield, Virginia 22161, USA.
6. H. Grieser, Impact on System Design of Cost Analysis of Specifications and Requirements, in Design to Cost and life Cycle Cost, North Atlantic Treaty Organization (NATO) Advisory Group for Aerospace Research and Development (AGARD) Conference Proceedings No. 289, May 1980, pp. 6.1–6.10. Available from the National Technical Information Service (NTIS), Springfield, Virginia 22161, USA.
7. W.M. Rowell, Design to Cost and the F-16 Multirole Fighter, in Design to Cost and Life Cycle Cost, North Atlantic Treaty Organization (NATO) Advisory Group for Aerospace Research and Development (AGARD) Conference Proceedings No. 289, May 1980, pp. 9.1–9.14. Available from the National Technical Information Service (NTIS), Springfield, Virginia 22161, USA.
8. K.E. Marks, H. Garrison-Massey, B.D. Bradley, An Appraisal of Models Used in Life Cycle Cost Estimation for United States Air Force (USAF) Aircraft Systems, Report No. R-2287-AF, October 1978. Prepared by the Rand Corporation, Santa Monica, California 90406, USA.

9. Logistics Support Cost Model User's Handbook, August 1976, Air Force Acquisition Logistics Division (AFLC), Headquarters, Washington, D.C.
10. Estimating Depot Maintenance Costs for Air Force Aircraft, AFLC Pamphlet 173-4, June 22, 1973. Prepared by Air Force Logistics Command, Department of the Air Force, Washington, D.C.
11. R.R. Fisher, et al., The Logistics Composite Model: An Overall View, Report No. RM-5544-PR, May 1968. Prepared by the Rand Corporation, Santa Monica, California 90406.
12. H.E. Boren, A Computer Model for Estimating Development and Procurement Costs of Aircraft (DAPCA-III), Report No. R-1854-PR, March 1976. Prepared by the Rand Corporation, Santa Monica, California 90406.
13. J.R. Nelson, F.S. Timson, Relating Technology to Acquisition Costs: Aircraft Turbine Engines, Report No. R-1288-PR, March 1974. Prepared by The Rand Corporation, Santa Monica, California.
14. J.P. Large, H.G. Campbell, D. Cates, Parametric Equations for Estimating Aircraft Airframe Costs, Report No. R-1693-1-PA&E, February 1976. Prepared by the Rand Corporation, Santa Monica, California.
15. Recoverable Inventory Control Using MOD-METRIC, Document No. AFLCP 57-13, Feb. 28, 1975, Prepared by Air Force Logistics Command (AFLC), Headquarters, Washington, D.C.
16. C.G. Sherbrooke, METRIC: A Multi-Echelon Technique for Recoverable Item Control, Report No. RM-5078-PR, November 1966. Prepared by the Rand Corporation, Santa Monica, California.
17. USAF Cost and Planning Factors, Report No. AFR 173-10, February 6, 1975. Department of the Air Force, Washington, D.C.
18. Air Force Manpower Standards, Report No. AFM 26-3, February 27, 1973. Department of the Air Force, Washington, D.C.
19. J.R. Nelson, An Approach to the Life-Cycle Analysis of Aircraft Turbine Engines, in the Application of Design to Cost and Life Cycle Cost to Aircraft Engines, North Atlantic Treaty Organization (NATO) Advisory Group for Aerospace Research and Development (AGARD) Lecture Series No. 107, May 1980, pp. 2.1–2.27. Available from the National Technical Information Service (NTIS), Springfield, Virginia 22161, USA.
20. E.J. Jones, Design to Life Cycle Cost Interaction of Engine and Aircraft, in the Application of Design to Cost and Life Cycle Cost to Aircraft Engines, North Atlantic Treaty Organization (NATO), Advisory Group for Aerospace Research and Development (AGARD) Lecture Series No. 107, May 1980, pp. 3.1–3.15. Available from the National Technical Information Service (NTIS), Springfield, Virginia 22161, USA.
21. R.F. Panella, M.A. Barga, R.G. McNally, The Role of Advanced Technology on Turbine Engine Life Cycle Cost, in Design to Cost and Life Cycle Cost, North Atlantic Treaty Organization (NATO) Advisory Group For Aerospace Research and Development (AGARD) Conference Proceedings No. 289, May 1980, pp. 26.1–26.13. Available from the National Technical Information Service (NTIS), Springfield, Virginia 22161, USA.

CHAPTER 11

Life Cycle Costing of Energy Systems and On-Surface Vehicles

11.1 Introduction

Nowadays, the life cycle costing concept is being applied in many diverse areas. The energy systems and on-surface vehicles are one of them. These are the two important areas where a large sum of money is being spent annually. A small percentage of savings in those areas can result in large sums of money. The concept of life cycle costing is becoming quite attractive in making decisions concerning energy systems and on-surface vehicles. A number of publications have appeared on both these topics in the published literature. These publications explore life cycle costing from several different directions.

This chapter is basically divided into two parts, i.e., life cycle costing of energy systems and life cycle costing of on-surface vehicles. Even though the basic intent of this chapter is to explore life cycle costing of energy systems and on-surface vehicles, the related cost topics are also covered. In some of the cases the related topics may or may not directly provide input to the life cycle cost. However, their knowledge will be useful for performing effective life cycle cost studies.

11.2 Life Cycle Costing of Energy Systems

Since the escalation of fuel prices an important consideration is being given to energy costs. For example, cost of energy consumption in buildings, in transportation systems, and in power generation. These types of cost directly or indirectly influence the life cycle cost of energy and related systems. This section specifically concentrates on topics related to energy.

11.2.1 Mean Useful Life of Energy Related Equipment

The information on average useful life of equipment is an important component in life cycle costing. This piece of information can be used in computing the field cost of the product. The average useful lives of selected energy related equipment are given below [1, 2]:

i) Steam turbines: 10–30 years
ii) Burners: 10 years
iii) Gasoline engines: 10 years
iv) Room air conditioners: 8 years
v) Oil fired furnaces: 10 years
vi) Diesel engines: 10–12 years
vii) Air compressors: 20 years
viii) Centrifugal pumps: 20–25 years
ix) Cooling towers (wood fill): 15–20 years
x) Electric motors: 20–25 years
xi) Fans (airfoil): 20 years
xii) Electric furnaces: 10 years
xiii) Transformers: 50 years
xiv) Gas turbines: 20 years
xv) Centrifugal fans: 25 years
xvi) Boilers: 20 years
xvii) Residential furnace (oil): 15 years
xviii) Residential furnace (gas): 15 years
xix) Residential furnace (electric): 15 years

11.2.2 Life Cycle Costing of Buildings

Since there are several types of buildings, one can take different approaches to life cycle cost. The usages of the building such as given below dictate the type of approach to be followed:

i) Apartments or a shopping center
ii) Offices
iii) School, university, or a hospital

For the purpose of performing life cycle cost analysis, one has to establish building's principal objective, general characteristics (e.g., wall construction, orientation), location, size, etc. However, the factors such as below have to be specifically defined for a life cycle study [2]:

i) Requirements concerned with ventilation
ii) Expected maintenance costs associated with equipment
iii) Location of the building
iv) Life expectancy of the building
v) Building U factors
vi) Building internal heat loads
vii) Mortgage debt (if applicable)
viii) Depreciation technique
ix) Interest rate
x) Income property tax rate (if applicable)
xi) Exterior design temperature for cooling, humidification, and heating
xii) Inflation factors (if applicable)
xiii) Insurance coverage rate (if applicable)
xiv) Utility rates for energy sources considered
xv) Building orientation
xvi) Building's weekly operating schedule
xvii) Interior design temperature for cooling, humidification, and heating
xviii) The weather year for the purpose of computing energy consumption
xix) Building's daily operating schedule

Life Cycle Cost Estimation Model I

This model was developed for the Department of Energy by the National Bureau of standards [3]. The building's design, operation, components, systems, and materials costs are taken into consideration in this model. The life cycle cost, LCC_B, is defined by

$$LCC_B = EC + RC + NRMC + SC + IC + OMC \qquad (11.1)$$

where

LCC_B is the life cycle cost
EC is the energy cost
RC is the replacement cost
$NRMC$ is the nonrecurring (non-fuel operating) maintenance cost
SC is the salvage cost
IC is the investment cost
OMC is the yearly recurring operation and maintenance cost (i.e., excluding fuel)

In the above equation, all the costs are at their present value. Each of the righthand side costs of Equation (11.1) is expressed below.

Energy Cost:

This is given by

$$EC = E_q \cdot P_u F_p \qquad (11.2)$$

where

E_q is the amount of purchased energy
P_u is the per unit cost of the purchased energy
F_p is the uniform present worth factor. This is modified for conditions such as fuel type, study period, and the appropriate region.

Replacement Cost:

This is expressed as follows:

$$RC = \sum_{i=1}^{k} C_R \cdot F_s \qquad (11.3)$$

where

k is the number of years
i is the year
C_R is the replacement cost
F_s is the single present worth factor

Non-Recurring Maintenance Cost:

This is defined by

$$NRMC = \sum_{i=1}^{k} C_{mn} \cdot F_s \qquad (11.4)$$

where C_{mn} is the annual non-recurring (non-fuel operating) maintenance cost.

Salvage Cost:

This cost is expressed as follows:

$$SC = \sum_{i=1}^{k} C_s \cdot F_s \qquad (11.5)$$

where C_s is the salvage cost

Investment Cost:

The cost is defined by

$$IC = C_R + C_I \cdot F_I \tag{11.6}$$

where

C_R is the repair or renovation cost (if applicable)
C_I is the investment cost
$F_I = 0.9$, is the investment adjustment factor; this factor is used to reflect estimated societal advantages (i.e. the ones are not satisfactorily, indicated by energy prices)

Recurring Operation and Maintenance Cost:

This is given by

$$OMC = F_{up} \cdot C_{om} \tag{11.7}$$

where

F_{up} is the uniform present worth factor
C_{om} is the yearly operating and maintenance costs (recurring and non fuel)

Life Cycle Cost Estimation Model II

This model was developed by the American Institute of Architects for facilities [4]. The four basic costs are considered in this model. These are cyclical renewal, project development, disposal, and use. Thus, the present value of life cycle cost of a facility is given by

$$LCC_f = \text{Recurring cost of the facility} + \text{nonrecurring cost of the facility} \tag{11.8}$$

One should note here that the righthand side costs of the above equation have to be expressed in their present values.

11.2.3 Energy Costing

This section presents three formulas concerned with energy cost estimation. Those formulas are as follows [5]:

Formula I

This formula is concerned with estimating electrical cost. This cost is composed of four components as below:

i) Fuel adjustment cost, C_{fa}
ii) Energy cost, C_e
iii) Power factor penalty cost, C_{pf}
iv) Demand cost, C_d

Thus, the electrical cost, EC, is as follows:

$$EC = C_{fa} + C_e + C_{pf} + C_d \tag{11.9}$$

Formula II

This formula is concerned with computing the yearly lighting cost. The lighting cost is defined by

$$C_L = \frac{P_L \cdot T_U \cdot Ph}{1{,}000} \tag{11.10}$$

where

C_L is the yearly lighting cost
P_L is the power consumed by the light bulb expressed in watts
T_u is the operational time (hours) of the bulb per year
P_h is the cost of the electricity per kilowatt hours, (\$/KWH)

Example 11.1

A 60-Watt light bulb is operated for 2,000 hours per year. The cost of the electricity is 8 cents per kilowatt hours. Calculate the yearly lighting cost.

In this example data for P_L, T_u, and P_h are specified. Thus, substituting the given data into Equation (11.10) yields

$$C_L = \frac{(60)(2{,}000)(0.08)}{1{,}000} = \$9.60$$

The annual lighting cost will be \$9.60.

Formula III

This formula is concerned with estimating cost to operate an A.C. motor. The cost to operate the motor is defined by

$$C_{om} = \frac{(0.746)H_p T_o P_h}{\theta} \tag{11.11}$$

where

C_{om} is the annual cost to operate the motor
H_p is the size of the motor (horsepower)

T_o is the annual operational time (hours) of the motor
P_h is the cost of the electricity per kilowatt hours, (\$/KWH)
θ is the motor efficiency

Example 11.2

The annual operational time of a 10-horsepower motor is 1,200 hours. The motor efficiency and the cost of electricity are 80% and 8 cents per kilowatt hour, respectively. Calculate the value of the annual cost to operate the A.C. motor.

Substituting the above data into Equation (11.11), we get

$$C_{om} = \frac{(0.746)(10)(1{,}200)(0.08)}{0.80}$$

$$= \$895.20$$

11.2.4 Life Cycle Cost Models for Energy Systems and Equipment

This section presents life cycle cost models of various energy related systems or equipment.

Life Cycle Cost Model I

This model was proposed by the Swedish [6] State Power Board (nuclear division) for the power plant productivity improvement. The life cycle cost is defined by

$$LCC_{pp} = C_I + C_{pr} + C_o + C_m \qquad (11.12)$$

where

LCC_{pp} is the life cycle cost
C_I is the investment cost
C_o is the operation cost
C_m is the cost of maintenance
C_{pr} is the cost of power replacement

The costs for design and construction, primary equipment, documentation and training, and initial investment in spares are included in the investment cost. The operation cost is composed of manpower cost, consumption of electric power cost, and other types of consumption costs. The costs for logistics, manpower, repetitive investments, and materials are the components of the maintenance cost. The power replacement cost is associated with the situation when the power plant is unavailable.

Life Cycle Cost Model II

This model is used to estimate life cycle cost of a room air conditioner. Thus, the life cycle cost of an air conditioner is given by

$$LCC_A = C_{AC} + C_{REA} \tag{11.13}$$

where

LCC_A is the life cycle cost of the room air conditioner
C_{AC} is the acquisition cost of the room air conditioner
C_{REA} is the recurring cost of the room air conditioner

Life Cycle Cost Model III

This model is for estimating the life cycle cost of a water heater. It is basically the same as the above model (for the room air conditioner). The water heater life cycle cost is as follows:

$$LCC_{wh} = C_{AW} + C_{REW} \tag{11.14}$$

where

LCC_{wh} is the life cycle cost of the water heater
C_{AW} is the acquisition cost of the water heater
C_{REW} is the recurring cost of the water heater

Life Cycle Cost Model IV

This model is concerned with estimating the life cycle cost of a high voltage circuit breaker. Generally, life cycle cost studies of high voltage power circuit breakers are not performed frequently because of lack of good data. For life cycle cost studies, the circuit breaker reliability related data needed are rate of force outages, scheduled unavailability, cost of scheduled maintenance, rate of forced outages that require a redundant breaker to operate, forced unavailability, and maintenance costs caused by failures. The life cycle cost [7] of a high voltage circuit breaker is defined by

$$LCC_{cb} = C_f + C_m + C_{fu} \tag{11.15}$$

where

LCC_{cb} is the life cycle cost of the circuit breaker
C_f is the fixed cost of the circuit breaker

C_m is the maintenance cost of the circuit breaker
C_{fu} is the cost associated with the unavailability of power transmission systems and power distribution systems

11.2.5 Cost Reliability Analysis

This section presents a cost reliability index. The purpose of presenting this index here is that it is related to life cycle costing. The index may simply be expressed [8] as a measure of the merit of an electric power distribution system design in terms of combined cost and reliability, and is given by

$$I_{CR} = [C_s \cdot I_{aid}]^{-1} \tag{11.16}$$

where

I_{CR} is the cost reliability index
C_s is the system cost per customer
I_{aid} is the system average interruption duration index

The high value of this index indicates good system design. More specifically, the greater the value of this index, the better the system design. The system average interruption duration index is defined by

$$I_{aid} = \frac{1}{k} \sum_{j=1}^{M} IN_{cj} \cdot T_j \tag{11.17}$$

where

k is the number of customers served
M is the number of annual interruptions
IN_{cj} is the number of customer interruptions
T_j is each interruption's length

For making the comparison of performance of electric power distribution systems, this is one of the most commonly used reliability indexes.

11.3 Life Cycle Costing of On-Surface Vehicles

Each year large sums of money are being spent on manufacture and operation of on-surface vehicles. Therefore, just like in energy systems, the cost effectiveness is also very important in vehicles. The life cycle costing is a mean for making effective decisions concerned with such systems. This section presents life cycle costing of civilian and military systems, separately.

11.3.1 Life Cycle Costing of Civilian Vehicles and Systems

This section presents four models [1, 9].

Urban Rail Life Cycle Cost Model

The life cycle cost of urban rail has two major components, i.e., system capital cost and system operating cost. Mathematically, it is as follows:

$$LCC_{UR} = C_{sc} + C_{so} \tag{11.18}$$

where

LCC_{UR} is the life cycle cost of urban rail
C_{sc} is the system capital cost
C_{so} is the system operating cost

The system capital cost is composed of cost components such as vehicles, signals, track, track work, stations, power substations and distribution, control, and maintenance facilities. Similarly, the cost elements of the system operating cost are the power, maintenance, transportation, etc. One should note here that this model represents the rail system rather than just a vehicle.

City Bus Transport Cost Model

The model is concerned with estimating city bus transport costs covering the total life period of vehicles. Thus, the overall cost over the life period is given by

$$C_0 = \sum_{i=1}^{12} C_i \tag{11.19}$$

where

C_1 is the cost of wages
C_2 is the cost of compulsory insurance
C_3 is the taxes
C_4 is the cost of fuel
C_5 is the cost of intermediate overhauls
C_6 is the cost of maintenance and checkup
C_7 is the vehicle purchasing price

C_8 is the cost of overheads
C_9 is the cost of general overhauls
C_{10} is the cost associated with current repairs
C_{11} is the cost of lubricants
C_{12} is the cost of tires

For more information on this model, the reader should consult Reference [9].

Automobile Life Cycle Cost Model

In this model, the total cost is broken down into several components. These are acquisition price, interest cost, maintenance and repair cost, cost of fuel, etc. The life cycle cost equation is as follows:

$$LCC_{AT} = C_A + C_{MR} + C_I + C_F + C_{ILF} + C_D - C_{SAL} \quad (11.20)$$

where

LCC_{AT} is the automobile life cycle cost
C_A is the automobile acquisition cost
C_{MR} is the cost of repair and maintenance
C_I is the interest cost
C_F is the fuel cost
C_{ILF} is the cost associated with insurance, license, and fees
C_D is the disposal cost (if applicable)
C_{SAL} is the salvage value (if applicable)

Other invisible costs (if any) can be added to appropriate costs of right hand side of Equation (11.20).

Cargo Ship Life Cycle Cost Model

In order to estimate the life cycle cost of a cargo ship, the total cost is normally broken down into four components: building costs, owner's costs, operating costs, and accommodation costs. The building cost includes the costs of machinery, hull, and outfitting. The elements of the owner's cost are the lawyer's fees, consulting fees, interest, naval architect's fees, and so on.

The operating cost has several cost components. Some of those are fuel, wages, cargo handling, maintenance, and insurance costs. The operating subsidiaries are also included in the operating costs. The costs associated with hull engineering, steel, and outfit form the accommodation cost. Thus, the cargo ship life cycle cost is given by

$$LCC_{cs} = C_o + C_{ow} + C_B + C_{AC} \quad (11.21)$$

where

LCC_{cs} is the life cycle cost of the cargo ship
C_o is the operating cost
C_{ow} is the owner's cost
C_B is the building cost
C_{AC} is the accommodation cost

11.3.2 Life Cycle Costing of Military Vehicles

Each year a large sum of money is being invested in military vehicles. A small percentage of saving in such investment can result in a large amount of money. The life cycle costing concept is a useful tool for making effective decisions regarding military vehicles. This section presents several aspects of costing in military vehicles.

Cost Breakdown Structure for the Combat Vehicle

The major components of the structure are shown in Figure 11.1. These are chassis, armament, fire control system, navigation and communication system, and turret. The elements of the chassis are suspension, roadwheels, engine, electrical system, fuel system, brakes, steering, auxiliary automotive systems, exhaust system, cooling system, etc. The elements such as gun, recoil, and buffers belong to armament. Some of the components of the fire control system are sensors, hydraulic power supply, laser range finder, and the ballistic computer. The navigation and communication system includes components such as inertial navigator, radio, and telephone. Finally, the elements such as turret structure, stowage, ventilation, and ammunition belong to turret.

Cost Analysis Methodology for Combat Vehicles

This section presents a basic methodology for combat vehicle support investment, and operating and support cost analysis [11]. The methodology is composed of a number of steps. Before we present these steps let us look into combat vehicle life cycle cost categories. The major ones are research and development, investment, and operating and support. There are several elements of these categories. For example, the components of operation and support are depot maintenance, deployed unit operations, personnel support and training, installations support, depot supply, etc.

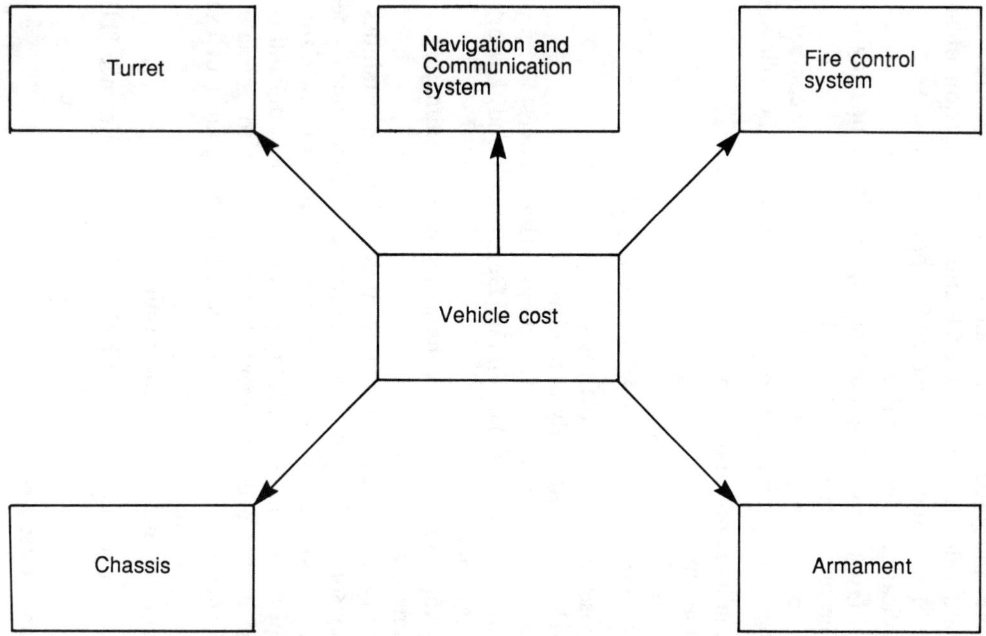

Figure 11.1. Cost breakdowns for the combat vehicle.

The cost analysis methodology is composed of ten steps as below:

i) Define and outline important issues in the predefense systems acquisition review council meetings. The pre-defense systems acquisition review council discussions should include topics such as the proposed system's description and characterization, ground rules for the cost evaluation; identification of support investment, and operating and support cost drivers, and the specification of alternatives.
ii) Identifying the existing reference system for comparing the costs of the proposed combat vehicle
iii) Develop system program definition statement for combat vehicle systems. The statement covers descriptions of the following:
    a) maintenance and support
    b) manning
    c) acquisition policies
    d) mission characteristics
iv) Choose relevant variable costs
v) Develop the cost analysis model
vi) Secure necessary data. The major data sources for proposed combat vehicles are the Military Department and the participating contractors.
vii) Develop conventions for consistency in cost estimating
viii) Estimating relevant costs
ix) Identify and recognize uncertainties. The uncertainties from many sources beset estimates of future combat vehicle operating and support costs. Some of these sources are operating and support environment, quality of data, technical difficulties experienced during development, techniques used to estimate costs, and changes in scope of the acquisition program (for example, costs, quantities).
x) Document and present end results.

All of the above steps are described in detail in Reference [11].

Maintenance Cost Drivers

There are a number of combat vehicle maintenance (unscheduled activities only) cost drivers. The major operating and support cost drivers for United States army combat vehicles were as follows [11]:

i) Track and suspension
ii) Electrical (automotive)

iii) Armament and fire control
iv) Engine

The percentage of total system failure: track and suspension, armament and fire control, engine, and electrical (automotive) accounted for 30%, 32%, 5%, and 20%, respectively. In addition, corresponding percentages of direct maintenance cost (parts and labor) were 48%, 35%, 10%, and 4%, respectively.

Major Contributors to Operating and Support Costs

There are several major historical contributors to operating and support costs. These contributors are associated with combat vehicle subsystems/components: drive sprocket wheel, track, oil and air filters, electrical harness, voltage generator and regulator, and final drive unit. The six major contributors are given below:

i) High maintenance requirement of drive sprocket wheel; the mean life of a drive sprocket wheel is 2,500 miles
ii) High mortality of track
iii) Time and mileage related replacement schedule for oil and air filters
iv) High maintenance associated with electrical harness
v) High mortality of voltage generator and regulator
vi) Oil leakage associated with the final drive unit

This leakage usually leads to internal failure.

Cost Estimation Models for Military Vehicles

This section presents two models to estimate cost of military vehicles. Equations for these models were developed in Reference [12] by examining cost data of military vehicles over the period of 30 years. The data cover vehicles such as army trucks, tanks, aircraft carriers, frigates, armoured personnel carriers, supersonic fighters, and transport aircrafts with installed power ranging from 200 to 300,000 horsepower.

Model I

This model is concerned with estimating mean unit cost of N number of vehicles. The mean unit cost, C_{mv}, of N vehicles built (in 1974 dollars) is given by

$$C_{mv} = 1{,}000T_{we} + 1{,}200HP_i(N)^{-0.33} \qquad (11.22)$$

where

HP_i is the installed power in horsepower (hp)
T_{we} is the empty weight expressed in tons (i.e., excluding fuel and payload)
N is the number of vehicles built

Model II

This model is concerned with estimating unit procurement cost in 1974 dollars of vehicles with large power/weight ratios and not produced in a mass quantity. In this case, the unit procurement cost in 1974 dollars is given by

$$C_{wmq} = 1{,}200HP_i(N)^{-0.33} \tag{11.23}$$

where

C_{wmq} is the unit procurement cost in 1974 dollars
HP_i is the installed power in horsepower (hp)
N is the number of vehicles built

Tank Life Cycle Cost Model

The life cycle cost of a tank is expressed by:

$$LCC_{TA} = C_{RD} + C_A + C_{OS} \tag{11.24}$$

where

LCC_{TA} is the life cycle cost of the tank
C_{RD} is the cost of research and development
C_A is the acquisition cost
C_{OS} is the operation and support cost

The acquisition cost is composed of two components as follows:

$$C_A = C_R + C_{NR} \tag{11.25}$$

where

C_R is the recurring cost
C_{NR} is the non-recurring cost

Similarly, the operation and support cost is given by

$$C_{os} = C_{Dm} + C_{mm} + C_{mp} + C_{is} + C_{ds} + C_{cn} \tag{11.26}$$

where

C_{Dm} is the cost associated with depot maintenance
C_{mm} is the cost of material modifications
C_{mp} is the cost associated with military personnel
C_{is} is the cost associated with indirect support
C_{ds} is the cost associated with other direct support
C_{cn} is the cost associated with consumption

11.4 Summary

This chapter is concerned with life cycle costing of energy systems and on-surface vehicles. Both these topics are presented separately. The mean useful life of a certain number of energy related equipment is given. The life cycle costing of buildings is discussed along with two life cycle cost models. Three formulas concerned with energy cost estimation are presented. These formulas deal with estimation of electrical cost, lighting cost, and cost to operate an A.C. motor. Mathematical models for estimating life cycle cost of a room air conditioner, a water heater, and a high voltage circuit breaker are presented. The cost-reliability index is defined.

Three models for estimating life cycle cost of the urban rail, automobile, and cargo ship are presented. A cost analysis methodology for combat vehicles is outlined along with the cost breakdown structure. Major operating and support cost drivers for military vehicles are listed. These are track and suspension, electrical (automotive), armament and fire control, and engine. A model to estimate life cycle cost of a military tank is presented.

11.5 Problems

1. What are the main cost components of the life cycle cost model developed for the Department of Energy by the National Bureau of Standards?
2. What are the similarities and differences between the life cycle cost models developed by the National Bureau of Standards and the American Institute of Architects.
3. What are the components of the electrical cost?
4. A 100-Watt light bulb is operated for 2,800 hours annually. The cost of the electricity is 10 cents per kilowatt hours. Compute the annual lighting cost.
5. The yearly operational time of a 20-horsepower A.C. motor is 2,500 hours. The motor efficiency is 85%. If the cost of electricity

is 10 cents per kilowatt hours, estimate the yearly cost to operate the motor.

References

1. M. Eddins-Earles, Factors, Formulas, and Structures for Life Cycle Costing, Published by Eddins-Earles, 89 Lee Drive, Concord, Mass. 01742, USA, 1981.
2. J.P. Manzo, Life-Cycle Costing—, Specifying Engineer, September 1976, pp. 71–78.
3. Life Cycle Cost Manual for The Federal Energy Management Program, National Bureau of Standards Handbook 135, United States Department of Commerce, Washington, D.C. December 1980.
4. Life Cycle Cost Analysis: A Guide for Architects, American Institute of Architects, 1735 New York Avenue, N.W., Washington, D.C., 1979.
5. R.J. Brown, R.R. Yanuck, Life Cycle Costing: A Practical Guide for Energy Managers, The Fairmont Press, Inc., P.O. Box 14227, Atlanta, Georgia 30324, 1980.
6. M. Anderson, T. Eriksson, L. Wrendenberg, V. Huusko, Life Cycle Cost in Management of Power Plant Availability Improvement, Proceedings of the International Reliability, Availability and Maintainability Conference for the Electric Power Industry, 1983, pp. 227–233.
7. C.R. Heising, Reliability and Maintenance Data Needed for High Voltage Circuit Breakers when Making Life Cycle Cost Studies, Proceedings of the International Reliability Conference for the Electric Power Industry, 1979, pp. 103–108.
8. N.E. Chang, Life-Cycle Cost Evaluation of Underground Distribution Systems, Specifying Engineer, October 1979, pp. 102–104.
9. F.H. Zalud, J. Lanc, Automobile Reliability—A Key to Lower Overall Transport Costs, Proceedings of the XIVth International Automobile Technical Congress of FISITA, London, June 25–30, 1972, pp. 6/25–6/32, Published by the Institution of Mechanical Engineers, London.
10. H. Benford, General Cargo Ship Economics & Design, Department of Naval Architecture and Marine Engineering, University of Michigan, Ann Arbor, August 1965.
11. M.R. Fiorello, L.G. Jones, Combat Vehicle System Operating and Support Costs: Guidelines for Analysis, Report No. LMI Task 75-1/3, Logistics Management Institute, 4701 Sangamore Road, Washington, D.C., 200 16, June 1977.
12. D.M. Dix, F.R. Riddell, Projecting Cost-Performance Trade-Offs for Military Vehicles, Astronautics and Aeronautics, Vol. 14, 1976, pp. 40–50.

CHAPTER 12

Introduction to Reliability

12.1 Introduction

Reliability is the probability that an item will carry out its mission satisfactorily for the desired period when used according to specified conditions. The history of the reliability discipline goes back to the years of World War II when Germans introduced the reliability concept to improve reliability of their rockets (i.e., V1 and V2). It was also the years of World War II when reliability was first recognized as a pressing need in the United States Armed Forces. During the period between 1945 and 1950 various studies related to reliability were conducted by Army, Air Force, and Navy. These studies revealed some startling results. In 1950, an ad hoc committee on reliability was established by the Department of Defense. Two years later in 1952, this committee became a permanent group known as the Advisory Group on the Reliability of Electronic Equipment (AGREE). A report was published by this group in 1957 [1]. In 1956, a book [2] entitled "Reliability Factors for Ground Electronic Equipment" was published by McGraw-Hill Book Company. Since the 1950's a vast amount of published literature on the subject has appeared and the field has developed into several specialized areas. Most of the specialized branches of reliability are discussed in References [3–6].

This chapter presents several introductory aspects of reliability.

12.2 Journals and Conferences Concerned with Reliability

There are several journals and conferences devoted partially or fully to the field of reliability. Most of the major journals are listed below:

i) IEEE Transactions on Reliability, jointly published by the IEEE and the American Society for Quality Control (ASQC)
ii) Microelectronics and Reliability: An International Journal, published by the Pergamon Press, U.K.

iii) Reliability Engineering and System Safety, published by the Elsevier Applied Science Publishers, U.K.
iv) Reliability Review, published by the American Society for Quality Control (ASQC)
v) Quality and Reliability Management, published by the MCB University Press, U.K.
vi) Transactions of the ASME: Journal of Vibration, Acoustics, Stress, and Reliability in Design, published by the American Society of Mechanical Engineers

Similarly, some of the major conferences concerned with reliability are as follows:

i) Annual Reliability and Maintainability Symposium
ii) Annual Inter-RAM (International Reliability, Availability and Maintainability), Conference for the Electronic Power Industry

12.3 Selected Definitions Used in Reliability Engineering

This section presents a number of selected definitions used in reliability engineering. These definitions are presented below.

Reliability: the probability that an item will carry out its mission satisfactorily for the desired period when used according to specified conditions.

Maintainability: the probability that a failed item will be restored to its satisfactory operational state within a specified total downtime when maintenance action is started according to stated conditions.

Redundancy: the existence of more than one means for carrying out a stated function.

Active Redundancy: the term used when all redundant items are operating simultaneously.

Downtime: the total time during which the product (or item) is not in a condition to carry out its stated function or mission.

Maintenance: all scheduled and unscheduled actions appropriate for keeping an item in a serviceable condition or restoring it to serviceability. It includes inspection, repair, remove and replace, servicing, testing, modification, etc.

Failure Rate: the number of failures of a product per unit measure of life (e.g., hours).

Failure: the termination of the ability of an item to carry out its stated mission or function.

12.4 Selected Mathematical Definitions

This section presents some selected mathematical definitions considered useful for reliability analysis.

12.4.1 Probability

The probability $P(E)$ of the outcome of event E is given by

$$P(E) = \lim_{k \to \infty} \left\{ \frac{N}{k} \right\} \tag{12.1}$$

In the above relationship, the event E occurs N times out of k repeated trials. In real life, the $P(E)$ is simply approximated from the following relationship:

$$P(E) = \frac{N}{k} \tag{12.2}$$

12.4.2 Probability of a Union of *k* Independent Events

This is expressed by

$$P(E_1 + E_2 + E_3 + \cdots + E_k) = 1 - \prod_{i=1}^{k} \{1 - P(E_i)\} \tag{12.3}$$

The union of events is denoted by the symbol $+$. $P(E_i)$ is the probability of event E_i; $i = 1, 2, 3, \ldots, k$.

12.4.3 Probability of an Intersection of *k* Independent Events

This is given by

$$P(E_1 \cap E_2 \cap E_3 \cap \cdots \cap E_k) = \prod_{i=1}^{k} P(E_i) \tag{12.4}$$

or

$$P(E_1 E_2 E_3 \cdots E_k) = P(E_1)P(E_2)P(E_3) \cdots P(E_k) \tag{12.5}$$

The intersection of events is denoted by the symbol $\cap$.

12.4.4 Cumulative Distribution Function, Expected Value, and Variance (Continuous Distribution)

The cumulative distribution function is defined by

$$F_c(t) = \int_0^t f(x)\,\mathrm{d}x \tag{12.6}$$

where

x is the continuous random variable
$F_c(t)$ is the cumulative distribution function
$f(x)$ is the probability density function (continuous)

The expected value (mean) and the variance are defined by

$$m = \int_0^\infty x f(x)\,\mathrm{d}x \tag{12.7}$$

and

$$\sigma^2 = \int_0^\infty (x - m)^2 f(x)\,\mathrm{d}x \tag{12.8}$$

where

m is the mean value
σ^2 is the variance

12.4.5 Laplace Transform

The Laplace transform of function $f(t)$ is defined by

$$F(s) = \int_0^\infty f(t)\,e^{-st}\,\mathrm{d}t \tag{12.9}$$

where

$f(t)$ is a function of variable t
s is the Laplace transform variable

Example 12.1

Find the Laplace transform of the following function:

$$f(t) = e^{-(\lambda + \mu)t} \tag{12.10}$$

where λ and μ are the constants and t the time variable. Substituting Equation (12.10) into Equation (12.9) yields

$$\begin{aligned} F(s) &= \int_0^\infty e^{-(\lambda+\mu)t} \cdot e^{-st}\,\mathrm{d}t \\ &= \int_0^\infty e^{-(s+\lambda+\mu)t}\,\mathrm{d}t \\ &= \left[\frac{e^{-(s+\lambda+\mu)t}}{-(s+\lambda+\mu)}\right]_0^\infty \\ &= \frac{1}{s+\lambda+\mu} \end{aligned} \tag{12.11}$$

12.4.6 Final-Value Theorem

If the following limits exist, then

$$\lim_{t \to \infty} f(t) = \lim_{s \to 0} sF(s) \tag{12.12}$$

The symbols t, s, $f(t)$, and $F(s)$ were defined earlier.

Example 12.2

If we have

$$f(t) = \frac{\mu_1}{\mu_1 + \mu_2}[1 - e^{-(\mu_1 + \mu_2)t}] \tag{12.13}$$

prove that the results obtained using left and right hand sides of Equation (12.12) are the same.

The symbols used in Equation (12.13) are defined below:

μ_1 and μ_2 are the constants
t is time variable

Substituting Equation (12.13) into left hand side of Equation (12.12) yields

$$\lim_{t \to \infty}\left[\frac{\mu_1}{\mu_1 + \mu_2}\{1 - e^{-(\mu_1 + \mu_2)t}\}\right] = \frac{\mu_1}{\mu_1 + \mu_2} \tag{12.14}$$

Using Equation (12.9), the Laplace transform of Equation (12.13) is

$$\begin{aligned} F(s) &= \int_0^\infty \frac{\mu_1}{\mu_1 + \mu_2}\{1 - e^{-(\mu_1 + \mu_2)t}\}\, e^{-st}\, dt \\ &= \frac{\mu_1}{\mu_1 + \mu_2}\left[\frac{e^{-st}}{-s} + \frac{e^{-(s + \mu_1 + \mu_2)t}}{(s + \mu_1 + \mu_2)}\right]_0^\infty \\ &= \frac{\mu_1}{\mu_1 + \mu_2}\left[\frac{1}{s} - \frac{\mu_1}{(s + \mu_1 + \mu_2)}\right] = \frac{\mu_1}{s(s + \mu_1 + \mu_2)} \end{aligned} \tag{12.15}$$

Substituting the above result into right hand side of Equation (12.12) yields

$$\lim_{s \to 0}\left[\frac{s\mu_1}{s(s + \mu_1 + \mu_2)}\right] = \frac{\mu_1}{\mu_1 + \mu_2} \tag{12.16}$$

It proves that results obtained using left and right hand sides of Equation (12.12) are the same.

12.5 Bathtub Hazard Rate Curve

This is a well known curve used to represent the failure behavior of many engineering items. The name of the curve is derived from its shape resembling bathtub as shown in Figure 12.1. As marked in the figure, the curve can be divided into three parts: burn-in period (decreasing hazard rate region), useful life period (constant hazard rate region, and wearout period (increasing hazard rate region). Failures occurring during the burn-in period are normally due to design or manufacturing defects. Figure 12.1 shows that as the "burn-in period" increases, the item's failures decrease until the starting of its useful life period. The "useful life period" failures are known as "random failures" because they occur unpredictably [5]. Some of the causes for the "useful life period" failures are undetectable defects, low safety factors, human errors, abuse, and higher random stress than expected. During the useful life period, the item's hazard rate remains constant and at the end of the useful life period, the hazard rate starts increasing. This increase is the result of item's wear-out. Some of the reasons for "wear-out period" failures are inadequate maintenance, wear due to aging, corrosion, creep, and incorrect

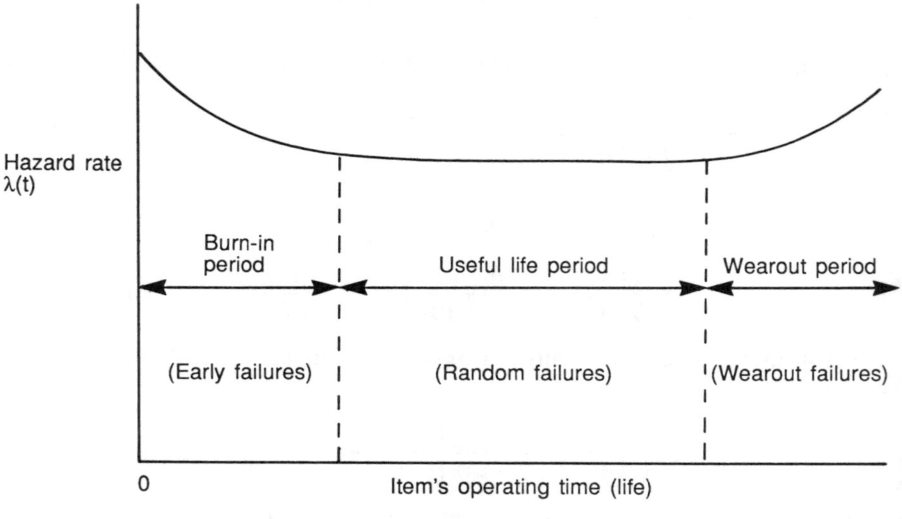

Figure 12.1. Bathtub hazard rate curve.

overhaul practices. Generally, in practice the reliability analysis considers only the middle portion of the curve shown in Figure 12.1. It means that the item's failure times are exponentially distributed.

12.6 General Formulas for Determining an Item's Reliability, Mean Time to Failure, and Hazard Rate

This section presents three general formulas to determine an item's reliability, mean time to failure, and hazard rate.

12.6.1 Reliability

Formula to determine an item's reliability is

$$R(t) = e^{-\int_0^t \lambda(x)\,\mathrm{d}x} \tag{12.17}$$

where

t is time
$\lambda(x)$ is the item's hazard rate or the time-dependent failure rate
$R(t)$ is the item's reliability

Example 12.3

Assume that the failure rate, λ, of an electric transformer is 0.007 failures per hour. Calculate the transformer's reliability for a 12-hour mission.

In this case, we have $\lambda(x) = \lambda = 0.007$ failures/hour, and $t = 12$ hours. Using these data in Equation (12.17) results in

$$\begin{aligned} R(12) &= e^{-\int_0^{12} (0.007)\,\mathrm{d}x} \\ &= e^{-[0.007x]_0^{12}} \\ &= e^{-(0.007)(12)} = 0.9194 \text{ or } 91.94\% \end{aligned}$$

Thus, the electric transformer's reliability is 0.9194.

12.6.2 Mean Time to Failure

The mean time to failure ($MTTF$) of an item is given by

$$MTTF = \int_0^{\infty} R(t)\,\mathrm{d}t \tag{12.18}$$

where

$R(t)$ is the item's reliability

Alternatively,

$$MTTF = \lim_{s \to 0} R(s) \tag{12.19}$$

where

s is the Laplace transform variable
$R(s)$ is the Laplace transform of the item's reliability

Example 12.4

An item's reliability is expressed by

$$R(t) = e^{-\lambda t} \tag{12.20}$$

where

$R(t)$ is the item's reliability
λ is the item's constant failure rate
t is time

Obtain an expression for mean time to failure using Equations (12.18) and (12.19). If the failure rate, λ, of the item is 0.007 failures/hour, calculate its mean time to failure.

Substitution of Equation (12.20) into Equation (12.18) yields

$$MTTF = \int_0^{\infty} e^{-\lambda t} \, dt = \frac{1}{\lambda} \tag{12.21}$$

The Laplace transform of Equation (12.20) is

$$R(s) = \frac{1}{s + \lambda} \tag{12.22}$$

Substituting Equation (12.22) in Equation (12.19) yields

$$MTTF = \lim_{s \to 0} \left[\frac{1}{s + \lambda} \right] = \frac{1}{\lambda} \tag{12.23}$$

This proves that Equations (12.18) and (12.19) yield the same result.

Subsituting the given data for λ in the right hand side of Equation (12.23) yields

$$MTTF = \frac{1}{(0.007)} = 142.857 \text{ hours}$$

The item's mean time to failure is 142.857 hours.

12.6.3 Hazard Rate

An item's hazard rate, $\lambda(t)$, can be obtained from the following relationships:

$$\lambda(t) = \frac{f(t)}{R(t)} \tag{12.24}$$

or

$$\lambda(t) = -\frac{1}{R(t)} \cdot \frac{dR(t)}{dt} \tag{12.25}$$

or

$$\lambda(t) = \frac{f(t)}{\int_t^{\infty} f(t)\, dt} \tag{12.26}$$

where

$f(t)$ is the failure density function
$R(t)$ is the reliability function
t is time

Example 12.5

Assume that as in Example 12.4 an item's reliability is expressed by

$$R(t) = e^{-\lambda t} \tag{12.27}$$

Obtain an expression for the item's hazard rate. Substitution of Equation (12.27) into Equation (12.25) results in

$$\begin{aligned} \lambda(t) &= -\frac{1}{e^{-\lambda t}} \cdot \frac{d(e^{-\lambda t})}{dt} \\ &= -\left[\frac{\lambda\, e^{-\lambda t}}{e^{-\lambda t}}\right] \\ &= \lambda \end{aligned} \tag{12.28}$$

In this case, the item's hazard rate is constant and it means its failure times are exponentially distributed.

12.7 Reliability Networks

This section is concerned with reliability evaluation of series, parallel, m-out-of-n, and standby arrangements [3, 5, 6].

12.7.1 Series Configuration

This type of arrangement is encountered quite frequently and is the simplest to analyze. The configuration represents a system with units in series. If anyone of the units fails, the system fails.

A k unit series system reliability is given by

$$R_{ss} = P(X_1 \cap X_2 \cap X_3 \cap \cdots \cap X_k] \tag{12.29}$$

where

R_{ss} is the series system reliability
$\cap$ is the symbol used to denote intersection of events
X_j is the event that unit k works normally (successfully)
k is the number of units in the series system
P is the probability

For statistically independent units, Equation (12.29) simplifies to

$$\begin{aligned} R_{ss} &= P(X_1)P(X_2)P(X_3) \cdots P(X_k) \\ &= R_1 R_2 R_3 \cdots R_k \end{aligned} \tag{12.30}$$

where $R_j = P(X_j)$, the unit j reliability; for $j = 1, 2, \ldots, k$. For the time dependent case, Equation (12.30) is rewritten to the following form:

$$R_{ss}(t) = R_1(t)R_2(t)R_3(t) \cdots R_k(t) \tag{12.31}$$

where t is time

If the failure times of units are exponentially distributed (i.e., units with constant failure rates), then the unit j reliability from Equation (12.20) is

$$R_j(t) = e^{-\lambda_j t} \tag{12.32}$$

where

$R_j(t)$ is the unit j reliability
λ_j is the constant failure rate of unit j

Substituting Equation (12.32) in Equation (12.31) results in

$$R_{ss}(t) = e^{-\lambda_1 t} e^{-\lambda_2 t} e^{-\lambda_3 t} \cdots e^{-\lambda_k t} \tag{12.33}$$

For identical units, Equation (12.33) becomes

$$R_{ss}(t) = e^{-k\lambda t} \tag{12.34}$$

where λ is the constant failure rate of a unit.

Example 12.6

A system has four independent units and all of the units must operate normally for the system success. Each unit reliability is 0.85. Calculate the system reliability.

In this case, we have $k = 4$ and $R_1 = R_2 = R_3 = R_4 = 0.85$. Thus, substituting these data into Equation (12.30) yields

$$R_{ss} = (0.85)(0.85)(0.85)(0.85)$$
$$= 0.5220$$

Thus, the series system reliability is 0.5220.

Example 12.7

A system has three independent units and all of the three units must operate normally for the system success. Failure rates of units 1, 2, and 3 are 0.002 failures/hour, 0.001 failures/hour, and 0.004 failures/hour, respectively. Calculate the system reliability for a 10-hour mission.

In this example, the data for following items are specified:

$$\lambda_1 = 0.002 \text{ failures/hour}$$
$$\lambda_2 = 0.001 \text{ failures/hour}$$
$$\lambda_3 = 0.004 \text{ failures/hour}$$
$$t = 10 \text{ hours}$$

Using these data in Equation (12.33) leads to

$$R_{ss}(10) = [e^{-(0.002)(10)}][e^{-(0.001)(10)}][e^{-(0.004)(10)}]$$
$$= 0.9324 \text{ or } 93.24\%$$

Thus, the system reliability is 0.9324 or 93.24%.

12.7.2 Parallel Configuration

This configuration represents a system with units in parallel. All units in the system operate simultaneously. The system will fail only, if all the units in the system fail. The unreliability of the system is expressed by

$$F_p = P[\bar{X}_1 \cap \bar{X}_2 \cap \bar{X}_3 \cap \cdots \cap \bar{X}_k] \tag{12.35}$$

where

F_p is the parallel system unreliability
$\bar{X}_j$ is the complementary event
k is the number of units in parallel

The remaining symbols were defined earlier in the chapter. For independent units, Equation (12.35) becomes

$$F_p = P(\bar{X}_1)P(\bar{X}_2)P(\bar{X}_3) \cdots P(\bar{X}_k)$$
$$= F_1F_2F_3 \cdots F_k \tag{12.36}$$

where

$F_j = P(\bar{X}_j)$, the unit j unreliability, for $j = 1, 2, 3, \ldots, k$

Since the sum of unit j reliability (R_j) and unreliability (F_j) is unity, Equation (12.36) is rewritten to the following form:

$$F_p = (1 - R_1)(1 - R_2)(1 - R_3) \ldots (1 - R_k) \tag{12.37}$$

where R_j is the unit j reliability, for $j = 1, 2, 3, \ldots, k$. Similarly, the parallel system reliability, R_p, is given by

$$R_p = 1 - F_p$$
$$= 1 - (1 - R_1)(1 - R_2)(1 - R_3) \cdots (1 - R_k) \tag{12.38}$$

As for the series configuration, when units' failure rates are constant, substituting Equation (12.32) in Equation (12.38) yields

$$R_p(t) = 1 - (1 - e^{-\lambda_1 t})(1 - e^{-\lambda_2 t})(1 - e^{-\lambda_3 t}) \cdots (1 - e^{-\lambda_k t})$$
$$= 1 - \prod_{j=1}^{k} (1 - e^{-\lambda_j t}) \tag{12.39}$$

where

t is time
$R_p(t)$ is the time dependent parallel system's reliability

For identical units Equation (12.39) reduces to

$$R_p(t) = 1 - (1 - e^{-\lambda t})^k \tag{12.40}$$

where λ is the constant failure rate of a unit

Example 12.8

A system has two active and independent units in parallel. Failure rates of units 1 and 2 are 0.005 failures/hour and 0.009 failures/hour, respectively. Calculate the system reliability for a 6-hour mission.

For $k = 2$, from Equation (12.39), we get

$$R_p(t) = 1 - (1 - e^{-\lambda_1 t})(1 - e^{-\lambda_2 t}) \tag{12.41}$$

Substituting the given data for λ_1, λ_2, and t in Equation (12.41) results in

$$R_p(6) = 1 - [1 - e^{-(0.005)(6)}][1 - e^{-(0.009)(6)}]$$
$$= 0.9984$$

Thus, the reliability of the two unit parallel system is 0.9984.

12.7.3 *m*-out-of-*n* System

In this case, n (active) units form a parallel system and at least m units must function successfully for the system success. In the case of independent and indentical units, the system reliability is given by

$$R_{m/n} = \sum_{j=m}^{n} \binom{n}{j} R^j (1 - R)^{n-j} \tag{12.42}$$

where

$$\binom{n}{j} = \frac{n!}{j!(n-j)!} \tag{12.43}$$

$R_{m/n}$ is the m-out-of-n system reliability
R is the unit's reliability

At $m = 1$, Equation (12.42) gives reliability for a parallel system and at $m = n$, it gives reliability for a series system.

If failure times of all units are exponentially distributed, then Equation (12.42) becomes

$$R_{m/n}(t) = \sum_{j=m}^{n} \binom{n}{j} [e^{-\lambda t}]^j [1 - e^{-\lambda t}]^{n-j} \tag{12.44}$$

where

t is time
λ is the constant failure rate of a unit
$R_{m/n}(t)$ is the m-out-of-n system reliability at time t

Example 12.9

A system is composed of four active and independent units. At least three units must function normally for the system success. Each unit reliability is 0.90. Compute the system reliability.

Substituting the given data into Equation (12.42) yields

$$R_{3/4} = \sum_{j=3}^{4} \binom{4}{j}(0.9)^j(1-0.9)^{4-j}$$

$$R_{3/4} = 4(0.9)^3 - 3(0.9)^4$$

$$= 0.9477$$

The system reliability is 0.9477 or 94.77%.

12.7.4 Standby System

In this case, one unit is operating and n units are on standby. In fact the system has $(n+1)$ units. The independent and identical unit standby system reliability is given by

$$R_{STD}(t) = e^{-\lambda t}\left[1 + \lambda t + \frac{(\lambda t)^2}{2!} + \cdots + \frac{(\lambda t)^n}{n!}\right] \tag{12.45}$$

where

t is time
$R_{STD}(t)$ is the standby system reliability
λ is constant failure rate of a unit
n is the number of units on standby

Equation (12.45) is subject to the following assumptions:

i) The standby units are as good as new
ii) The switching arrangement is perfect
iii) The failure times of all units are exponentially distributed

Example 12.10

A standby system is composed of three independent and identical units (i.e., one operating and two on standby). Each unit's failure rate is 0.004 failures/hour. Calculate the system reliability for a 150-hour mission using Equation (12.45).

In this example, we have $\lambda = 0.004$ failures/hour, $t = 150$ hours, and $n = 2$. Thus, substituting these data into Equation (12.45) results in

$$R_{STD}(150) = e^{-(0.004)(150)}\left[1 + (0.004)(150) + \frac{[(0.004)(150)]^2}{2}\right]$$
$$= (0.5488)[1 + 0.60 + 0.18]$$
$$= 0.9769$$

Thus, the standby system reliability is 0.9769 or 97.69%.

12.8 Mean Time to Failure (*MTTF*) of Reliability Systems

This section presents formulas to determine mean time to failure (*MTTF*) of series, parallel, *m*-out-of-*n*, and standby systems when units' failure are exponentially distributed.

12.8.1 Series System

Substituting Equation (12.33) into Equation (12.18) leads to

$$MTTF_{ss} = \int_0^\infty [e^{-\lambda_1 t} \cdot e^{-\lambda_2 t} \cdot e^{-\lambda_3 t} \cdots e^{-\lambda_k t}]\, dt$$
$$= \int_0^\infty e^{-\sum_{j=1}^{k} \lambda_j t}\, dt$$
$$= \left[\frac{e^{-\sum_{j=1}^{k} \lambda_j t}}{-\sum_{j=1}^{k} \lambda_j}\right]_0^\infty$$
$$= \frac{1}{\sum_{j=1}^{k} \lambda_j} \tag{12.46}$$

where

$MTTF_{ss}$ is the mean time to failure of the series system
λ_j is the constant failure rate of unit j; for $i = 1, 2, \ldots, k$

For identical units, Equation (12.46) simplifies to

$$MTTF_{ss} = \frac{1}{k\lambda} \tag{12.47}$$

where λ is the constant failure rate of a single unit

Example 12.11

A series system is composed of six independent units. Failure rates of units 1, 2, 3, 4, 5, and 6 are 0.0004 failures/hour, 0.0005 failures/hour, 0.0006 failures/hour, 0.0007 failures/hour, 0.0008 failures/hour, and 0.0009 failures/hour, respectively. Determine the system mean time to failure.

Substituting these data into Equation (12.46) leads to

$$\begin{aligned} MTTF &= (\lambda_1 + \lambda_2 + \lambda_3 + \lambda_4 + \lambda_5 + \lambda_6)^{-1} \\ &= [(0.0004) + (0.0005) + (0.0006) + (0.0007) + (0.0008) \\ &\quad + (0.0009)]^{-1} \\ &= 256.41 \text{ hours} \end{aligned}$$

Thus, the mean time to failure of the series system is 256.41 hours.

12.8.2 Parallel System

Substituting Equation (12.39) into Equation (12.18) yields

$$\begin{aligned} MTTF_{ps} &= \int_0^{\infty} \left[1 - \prod_{j=1}^{k} (1 - e^{-\lambda_j t}) \right] \mathrm{d}t \\ &= \sum_{j=1}^{k} \frac{1}{\lambda_j} - \sum_{j=1}^{k-1} \sum_{i=j+1}^{k} \frac{1}{\lambda_j + \lambda_i} + \cdots + (-1)^{k-1} \left[\sum_{j=1}^{k} \lambda_j \right]^{-1} \end{aligned} \tag{12.48}$$

where $MTTF_{ps}$ is the mean time to failure of the parallel system
For independent and identical units, the parallel system mean time to failure is given by

$$MTTF_{ps} = \frac{1}{\lambda} \sum_{j=1}^{k} \frac{1}{j} \tag{12.49}$$

where

k is the number of units in the parallel system
λ is the constant failure rate of a unit

Example 12.2

Assume that in Equation (12.48), we have $k = 2$, $\lambda_1 = 0.0002$ failures/hour, and $\lambda_2 = 0.0004$ failures/hour. Calculate the parallel system mean time to failure.

Using these data in Equation (12.48) yields

$$MTTF_{ps} = \frac{1}{\lambda_1} + \frac{1}{\lambda_2} - \frac{1}{\lambda_1 + \lambda_2}$$

$$= \frac{1}{(0.0002)} + \frac{1}{(0.0004)} - \frac{1}{(0.0002) + (0.0004)}$$

$$= 5833.33 \text{ hours}$$

The mean time to failure of the two unit parallel system is 5833.33 hours.

12.8.3 *m*-out-of-*n* System

The mean time to failure of the *m*-out-of-*n* system is obtained by substituting Equation (12.44) in Equation (12.18) as follows:

$$MTTF_{m/n} = \int_0^\infty \sum_{j=m}^{n} \binom{n}{j} [e^{-\lambda t}]^j [1 - e^{-\lambda t}]^{n-j} \, dt$$

$$= \frac{1}{\lambda} \sum_{j=m}^{n} \frac{1}{j} \qquad (12.50)$$

where $MTTF_{m/n}$ is the mean time to failure of the *m*-out-of-*n* system

Example 12.3

A system is composed of four independent and identical active units. At least two of the units must work normally for the mission success. Each unit's failure rate is 0.002 failures/hour. Calculate the system mean time to failure.

In this example, we have $\lambda = 0.002$ failures/hour, $m = 2$, and $n = 4$. Using these data in Equation (12.50) results in

$$MTTF_{2/4} = \frac{1}{\lambda} \sum_{j=2}^{4} \frac{1}{j}$$

$$= \frac{1}{\lambda} \left[\frac{1}{2} + \frac{1}{3} + \frac{1}{4} \right]$$

$$= \frac{13}{12\lambda} = \frac{13}{12(0.002)} = 541.667 \text{ hours}$$

The system mean time to failure is 541.667 hours.

12.8.4 Standby System

In order to obtain mean time to failure of a standby system we substitute Equation (12.45) in Equation (12.18), as follows:

$$MTTF_{ST} = \int_0^\infty e^{-\lambda t}\left[1 + \lambda t + \frac{(\lambda t)^2}{2!} + \cdots + \frac{(\lambda t)^n}{n!}\right] \mathrm{d}t$$
$$= \frac{(n+1)}{\lambda} \qquad (12.51)$$

where $MTTF_{ST}$ is the mean time to failure of the standby system

Example 12.4

Assume that in Equation (12.51) $n = 1$, and $\lambda = 0.0002$ failures/hour. Compute the standby system mean time to failure.

Thus, from Equation (12.51), we get

$$MTTF_{ST} = \frac{1+1}{(0.0002)} = 10{,}000 \text{ hours}$$

The mean time to failure of the standby system is 10,000 hours.

12.9 Failure Rate of a Series System

In order to obtain failure rate of an independent unit series system, we substitute Equation (12.33) in Equation (12.25) to get

$$\lambda_{ss}(t) = -\frac{1}{e^{-\sum_{j=1}^{k} \lambda_j t}} \cdot \frac{\mathrm{d}\left[e^{-\sum_{j=1}^{k} \lambda_j t}\right]}{\mathrm{d}t}$$
$$= -\frac{1}{e^{-\sum_{j=1}^{k} \lambda_j t}} \left[\left(-\sum_{j=1}^{k} \lambda_j\right) e^{-\sum_{j=1}^{k} \lambda_j t}\right]$$
$$= \sum_{j=1}^{k} \lambda_j \qquad (12.52)$$

where

λ_j is the constant failure rate of unit j
$\lambda_{ss}(t)$ is the failure rate of the series system

Table 12.1. Failure rates of selected items.

Serial no.	Item	Failure rate (failures/10^6 hours)
1	General A.C. motors (under 5 hp)	1.20–28.4
2	General A.C. motors (5–200 hp)	1.20–58.4
3	Circuit brakers (general, less than 33 kV)	2.0
4	Diesel engine (with control, fuel oil, cooling, starting, and lube subsystems)	30.3 to 1733
5	Power distribution transformer (all types, all sizes)	0.10–6.2
6	Ball bearings (heavy duty)	20.0
7	Valves (all types)	0.215–91.2
8	A.C. generators (general)	7.0
9	Batteries (thermal)	94.3*
10	Pumps	200*
11	Pressure vessels (general)	3.0

* Failures per billion hours.

Equation (12.52) is subject to the assumption that the failure times of each unit are exponentially distributed. Failure rates of selected items are presented in Table 12.1.

Example 12.15

A system is composed of four independent subsystems. If anyone of the subsystems fails, the system fails. Failure rates of subsystems 1, 2, 3, and 4 are 0.001 failures/hour, 0.002 failures/hour, 0.003 failures/hour, and 0.004 failures/hour, respectively. Calculate the system failure rate.

Substituting the specified data into Equation (12.52), we get

$$\begin{aligned}\lambda_{ss} &= \lambda_1 + \lambda_2 + \lambda_3 + \lambda_4 \\ &= (0.001) + (0.002) + (0.003) + (0.004) \\ &= 0.01 \text{ failures/hour}\end{aligned}$$

Thus, the failure rate of the four-unit series system is 0.01 failures/hour.

12.10 Maintainability Measures

This section is concerned with various aspects of maintainability. The maintainability function is defined by

$$MA(t) = \int_0^t f_r(t)\,\mathrm{d}t \tag{12.53}$$

where

t is time
$f_r(t)$ is the probability density function of repair times
$MA(t)$ is known as the maintainability function. Alternatively, it is the probability that repair will be accomplished in time t when it begins at time t equal to zero

The mean time to repair ($MTTR$) is given by

$$MTTR = \int_0^\infty [1 - MA(t)]\,\mathrm{d}t \tag{12.54}$$

Example 12.16

Assume that the repair times of a piece of equipment are exponentially distributed. Obtain an expression for the maintainability function. The repair time probability density function is defined by

$$f_r(t) = \frac{1}{MTTR}\,e^{-\left[\frac{1}{MTTR}\right]t} \tag{12.55}$$

where

$f_r(t)$ is the repair time probability density function
t is time
$MTTR$ is the mean time to repair

Using Equation (12.55) in Equation (12.53) yields

$$\begin{aligned} MA(t) &= \int_0^t \left\{\frac{1}{MTTR}\,e^{-\left[\frac{1}{MTTR}\right]t}\right\}\mathrm{d}t \\ &= \frac{1}{MTTR}\left[-(MTTR)\,e^{-\left[\frac{1}{MTTR}\right]t}\right]_0^t \\ &= \frac{1}{MTTR}\left[-(MTTR)\,e^{-\left[\frac{1}{MTTR}\right]t} + (MTTR)\right] \\ &= 1 - e^{-\left[\frac{1}{MTTR}\right]t} \end{aligned} \tag{12.56}$$

The instantaneous repair rate is defined by

$$\mu(t) = \frac{f_r(t)}{1 - MA(t)} \tag{12.57}$$

where $\mu(t)$ is the instantaneous repair rate

Example 12.17

Assume that the repair time probability density function of a system is given by Equation (12.55). Obtain an expression for the system repair rate.

Substituting Equations (12.55) and (12.56) into Equation (12.57) yields

$$\mu(t) = \frac{\left(\frac{1}{MTTR}\right) e^{-\left[\frac{1}{MTTR}\right]t}}{1 - \left[1 - e^{-\left[\frac{1}{MTTR}\right]t}\right]} = \frac{1}{MTTR} \tag{12.58}$$

12.10.1 Mean Time to Repair of a System

In the case of exponentially distributed repair times, the system mean time to repair ($MTTR_s$) is

$$MTTR_s = \frac{\sum_{j=1}^{k} \lambda_j T_j}{\lambda_s} \tag{12.59}$$

where

λ_s is the failure rate of the entire system
k is the number of repairable parts
λ_j is the component j failure rate
T_j is the system repair time when component j fails

12.10.2 Average Time Between Maintenance

The average time between maintenance ($ATBM$) is given by

$$ATBM = \left[\frac{1}{ACMT} + \frac{1}{APMT}\right]^{-1} \tag{12.60}$$

where

$ACMT$ is the average corrective maintenance time interval
$APMT$ is the average preventive maintenance time interval

12.11 Availability and Unavailability

Availability and unavailability of an item are expressed [3, 6] by

$$AV(t) = \frac{1}{(\lambda + \mu)} [\mu + \lambda e^{-(\lambda + \mu)t}] \tag{12.61}$$

and

$$UAV(t) = \frac{\lambda}{\lambda + \mu} [1 - e^{-(\lambda + \mu)t}] \tag{12.62}$$

where

$AV(t)$ is the item's availability
$UAV(t)$ is the item's unavailability
λ is the constant failure rate of the item
μ is the constant repair rate of the item
t is time

As t becomes very large, Equations (12.61) and (12.62) reduce to

$$AV = \frac{\mu}{\lambda + \mu} = \frac{MTTF}{MTTF + MTTR} \tag{12.63}$$

and

$$UAV = \frac{\lambda}{\lambda + \mu} = \frac{MTTR}{MTTR + MTTF} \tag{12.64}$$

where

AV is the steady state availability
UAV is the steady state unavailability
$MTTR$ is the mean time to repair
$MTTF$ is the mean time to failure

Example 12.18

Assume that failure and repair times of a system are exponentially distributed. The system mean time to failure ($MTTF$) and mean time to repair ($MTTF$) are 2,000 and 10 hours, respectively. Calculate the system steady state availability and unavailability.

In this example, we have $MTTR = 10$ hours, and $MTTF = 2{,}000$ hours. Using these data in Equations (12.63) and (12.64), we get

$$AV = \frac{2{,}000}{2{,}000 + 10} = 0.995$$

and

$$UAV = \frac{10}{2{,}000 + 10} = 0.005$$

Thus, the system availability and unavailability are 0.995 and 0.005, respectively.

12.12 Summary

This chapter presented several introductory aspects of engineering reliability. The history of reliability is briefly discussed. Major journals and conferences concerned with reliability and maintainability are listed. A number of selected definitions used in reliability engineering is presented. The bathtub hazard rate curve is described. General formulas for determining an item's reliability, mean time to failure, and hazard rate are presented. Equations are presented to calculate reliability and mean time to failure of series, parallel, m-out-of-n, and standby systems. An expression is developed for determining failure rate of a series system. Failure rates of selected items are presented. The maintainability function is defined. Formulas to determine mean time to repair and instantaneous repair rate are given along with item availability and unavailability expressions.

12.13 Problems

1. Write a 300-word essay on the history of reliability.
2. Define the following terms:
   i) Availability
   ii) Unavailability
   iii) Failure rate
   iv) Repair rate
3. Find the Laplace transform of the following function:

$$f(t) = \frac{\lambda}{\lambda + \mu}[1 - e^{-(\lambda + \mu)t}] \qquad (12.65)$$

where

t is time
λ is the constant failure rate
μ is the constant repair rate

4. What are the reasons for the product "burn-in period" failures?
5. Prove that the reliability of a series system is given by

$$R_{ss}(t) = e^{-\sum_{i=1}^{n} \lambda_i t} \tag{12.66}$$

where

$R_{ss}(t)$ is the series system reliability
λ_i is the constant failure rate of unit i
t is time
n is the number of units

6. A system is composed of four active and independent units in parallel. The failure rate of each unit is 0.0007 failures per hour. Calculate the system reliability for a 100-hour mission.
7. A system has three active and independent units. At least two of the units must function normally for the system success. The reliability of each unit is 0.95. Calculate the system reliability.
8. A standby system has four independent units (i.e., one operating and three on standby). The failure rate of each unit is 0.005 failures/hour. Calculate the system mean time to failure.
9. Assume that the repair times of a system are Weibull distributed. Obtain an expression for the system maintainability function.
10. Failure and repair rates of a system are 0.0005 failures per hour and 0.09 repairs/hour, respectively. Calculate the system steady state unavailability.

References

1. Reliability of Military Electronic Equipment, Report of the Advisory Group on Reliability of Electronic Equipment (AGREE), Office of the Assistant Secretary of Defense, Washington, D.C., June 4, 1957.
2. K. Henney, et al, Editors, Reliability Factors for Ground Electronic Equipment, McGraw-Hill Book Company, New York, 1956.
3. B.S. Dhillon, Reliability Engineering in Systems Design and Operation, Van Nostrand Reinhold Company, New York, 1983.
4. B.S. Dhillon, C. Singh, Engineering Reliability: New Techniques and Applications, John Wiley & Sons, New York, 1981.
6. B.S. Dhillon, Mechanical Reliability: Theory, Models and Applications, American Institute of Aeronautics and Astronautics, Washington, D.C., 1988.

APPENDIX

Bibliography of Literature on Life Cycle Costing and Related Areas

A.1 Introduction

Over the past decades a large number of publications have appeared on life cycle costing and related areas. This appendix presents selective references grouped into several categories: life cycle costing, reliability improvement warranty, software costs, manufacturing costs, quality costs, cost estimation, maintenance costs, operating costs and miscellaneous. Some of the references listed in each of those categories may not be totally relevant but are included because of their some relevance to the topic. In addition, there could be some overlap of references included in each category. These references will serve the reader well in the event he or she wishes to probe deeper into the area of his or her interest. Sections A.2, A.3, A.4, A.5, A.6, A.7, A.8, and A.9 present references on life cycle costing, reliability improvement warranty, software costs, manufacturing costs, quality costs, cost estimation, maintenance costs, and operating costs, respectively. The miscellaneous publications are listed in section A.10.

A.2 Life Cycle Costing*

1. R.W. Ackerman, "Introduction to the Application of Design to Cost and Life Cycle Cost To Aircraft Engines", in The Application of Design to Cost and Life Cycle Cost to Aircraft Engines, AGARD Lecture Series No. 107, May 1980, pp. 1-1 to 1-2. Available from National Technical Information Service (NTIS), Springfield, Virginia, U.S.A.
2. J.F. Adams, W.L. Massey, Jr., and M. Dmytryszyn, "Computer Speeds Economic Evaluations", in Cost Engineering in the Process Industries, Edited by Cecil H. Chilton and the Staff of Chemical Engineering, McGraw-Hill Book Company, Inc., New York, 1960, pp. 433–438.

---

* Almost half of the references listed in this section were taken (with permission from MCB University Press) from an article by Y. Gupta and W. S. Chow, "Twenty-Five Years of Life Cycle Costing—Theory and Applications: A Survey", The International Journal of Quality & Reliability Management, Vol. 2, 1985, pp. 51–76.

3. R.E. Adel, "Life Cycle Cost Model for Inertial Systems", Northrop Corporation, Electronics Division, M and ILS, Hawthorne, California, U.S.A., June 1974.
4. Advisory Group for Aerospace Research and Development (AGARD), "Methodology for Control of LCC for Avionic Systems", AGARD LS-100, April 1979, Available from National Technical Information Service (NTIS), 5285 Port Royal Road, Springfield, Virginia 22161, U.S.A.
5. Air Conditioning, Heating, and Refrigeration News, "BRI Workshop Hears of Life Cycle Costing Project by NBS", Vol. 122, January 1975, p. 5.
6. Air Force Logistics Command, "Life Cycle Cost Reference Notebook", Prepared by The Boeing Aerospace Company, Seattle, USA, August 1973.
7. R.C. Aikman, R.M. Schwartz, Life Cycle Cost Analysis Handbook: A Manual for Health Planners, ABT Publications, Cambridge, Massachusetts, USA, 1976.
8. P. Albrecht, "Economic Evaluation of Step-up Transformer Alternatives", Presented to Edison Power Electric Research Institute, Electrical System and Equipment Committee, February 22, 1978, USA.
9. H.C. Alexander, "A Guide to Estimating Aircraft Logistics Support Costs During Source Selection Evaluation", Cost Analysis Paper No. MCCC-70-002, Air Force Logistics Command, Ohio, USA, May 1970.
10. American Institute of Architects, "Life Cycle Cost Analysis, A Guide for Architects", American Institute of Architects, 1977.
11. R.H. Anderson, "Models and Methodology for Life Cycle Cost Test and Evaluation Analyses", Report No. OAS-TR-73-6, Air Force Systems Command, U.S.A., July 1973.
12. J.E. Anderson, "Cost Effectiveness", in Transit Systems Theory by J. Edward Anderson, Lexington Books, D.C. Heath and Company, Lexington, Massachusetts, USA, 1978, pp. 97–109.
13. J.E. Anderson, "Life Cycle Cost and Reliability Allocation", in Transit Systems Theory by J. Edward Anderson, Lexington Books, D.C. Heath and Company, Lexington, Massachusetts, USA, 1978, pp. 203–220.
14. J.E. Anderson, "Life-cycle Costs and Reliability Allocation in Automated Transit Systems", High Speed Ground Transportation Journal, Vol. 11, No. 1, Spring 1977, pp. 1–18.
15. M. Anderson, T. Eriksson, L. Wrenderberb, V. Huusko, "Life Cycle Cost in Management of Power Plant Availability Improvement", Proceedings of the 1983 International Reliability, Availability, Maintainability Conference for Electric Power Industry, 1983, pp. 227–233.
16. R. H. Anderson, T.E. Dixon, "Design to Cost Models: Helping Program Managers Manage Programs", Defense Mangement Journal, Vol. 12, No. 1, January 1976, pp. 65–71.

17. A. Anon, Life Cycle Cost Analysis of Electric Lighting Systems", Light Design Applications, Vol. 10, No. 5, May 1980, pp. 43–48.
18. J.R. Apel and D.E. Wiltse, "Applying Design-to-Life Cycle Cost Methods During Engine Advanced Development," Society of Automotive Engineers, Aerospace Meeting, San Diego, California, November 27–30, 1978, 10 pages. SAE Paper No. 781030.
19. J.A. Arceri, E.T. Parascos, "Economics of Reliability Purchasing Distribution Equipment with the Lowest Life Cycle Cost", Proceedings of the 1978 Reliability Conference for the Electric Power Industry, 1978, pp. 104–110.
20. J.S. Arendt, M.L. Casada and D.K. Lorenzo, "Economic Risk Assessment of a Petroleum Refinery", Proceedings of the Annual Reliability and Maintainability Symposium, 1985, pp. 329–335.
21. N.R. Augustine, "Is Life Cycle Cost Costing Lives?", Armed Forces Journal International, February 1978, pp. 110–115.
22. G. Auton, R.S. Howard, D. Parkes, H.W. Turner and H.S. Wood, "A Study of the Life-Cycle Costs of Vacuum Switchgear in Distribution Systems", Fourth International Conference on Electricity Distribution (CIRED), May 23–27, 1977, London, England.
23. R.H. Avery, "Life Cycle Costing of High-Efficiency Air Filter Plants", Plant Engineering, Vol. 31, No. 18, September 1, 1977, pp. 80–83.
24. D.W. Baggs, "Earth Covered: Low, Life-cycle-cost Housing", Architectural Science Review, Vol. 26, No. 3–4, September-December 1983, pp. 108–111.
25. R.L. Baglow, "The Reliability Parameter and Its Importance for Life Cycle Management," Microelectronics and Reliability, Vol. 14, 1975, pp. 91–105.
26. B. Bailey, M. Graves, E. Szten, "Reference Appendixes for Life Cycle Costing Procurement Techniques", Report No. UMTA-VA-06-0112-84-2, December 1984, 240 pp. Available from U.S. Department of Commerce, National Technical Information Service, Springfield, Virginia, U.S.A.
27. W.R. Baker, "Life Cycle Costing of Energy Systems", Proceedings of the 4th Annual UMR-DNR (University of Missouri-Rolla, Missouri, Department of Natural Resources) Conference on Energy, University of Missouri-Rolla, October 1977, Published by University of Missouri-Rolla, External Division, 1978, pp. 108–114.
28. S. Banerjee, and L. Waverman, "Life Cycle Costs of Coal and Nuclear Generating Stations", Report No. 36, July 1978, A Report prepared for Royal Commission on Electric Power Planning, Ontario Department of Energy, Toronto, Ontario, Canada.
29. H.S. Baras, "The Use of Life-Cycle Cost Estimates in Weapon System Acquisitions", Industrialization Forum, Vol. 6, Number 3–4, 1975, pp. 33–36.

30. R.K. Barasia and T.D. Kiang, "Development of a Life Cycle Management Cost Model", Proceedings of the Annual Reliability and Maintainability Symposium, 1978, pp. 254–259. Available from IEEE, New York, U.S.A.
31. R.K. Barasia, "Mathematical Model-Development of a Life Cycle Management Cost Model", Report No. 5J 30-4082-5, Bell-Northern Research, Published under DSS contract No. OSR 5-0102 for the Department of National Defense, Ottawa, Canada, December 1977.
32. R.K. Barasia, C.R. Brook and BNR LCC Project Team, "Development of a Life Cycle Management Cost Model (LCC)", Report No. 5J 30-4082-4, March 1978, Reliability Engineering Technology Division, Bell Northern Research Ltd., Ottawa, Ontario, Canada.
33. R.K. Barasia, C.R. Brook and R.B. Williams, "Methodology-Development of a Life Cycle Management Cost Model", Report No. 5 J 30-4082-2, Bell-Northern Research, Published Under DSS contract No. OSR5-0102 for the Department of National Defence, Ottawa, Canada, April 1977.
34. R.K. Barasia and C.R. Brook, "Case studies-Development of a Life Cycle Management Cost Model", Report No. 5J 30-4082-3, Bell-Northern Research, Published Under DSS Contract No. OSR5-0102 for the Department of National Defence, Ottawa, Canada, March 1978.
35. D.A. Baron, "A Logistics Life Cycle Support Cost Model", Report No. SLSR-5-69, The Air Force Institute of Technology, Ohio, August 1969, U.S.A.
36. L. Barrett, and E. Waddell, "Cost of Air Pollution Damage: A Status Report", A Report Written for Environmental Protection Agency, USA, May 1975. Available from Environmental Protection Agency, Washington, D.C., U.S.A.
37. I. Bazovsky, Jr., "Life Cycle Profits Instead of Life Cycle Costs", Proceedings of Annual Reliability and Maintainability Symposium, 1974, pp. 377–382.
38. R.K. Beach, "Life-Cycle Cost Equations", Building Research Note No. 150, September 1979, Available from Division of Building Research, National Research Council of Canada, Ottawa, Canada.
39. J.A. Belding, "Energy Conservation and Life Cycle Costing Methods", Energy, Vol. 3, No. 14, 1978, pp. 421–426.
40. C.F. Bell, "Reliability of Aircraft as Determined by Operational Field Tests: The Need for Proper Test Design and Data Requirements", Report No. P-4054, The Rand Corporation, Santa Monica, California, April 1969.
41. C.F. Bell, Determining Economic Quantities of Maintenance Resources: A Minuteman Application", Report No. RM-3308-PR, The Rand Corporation, Santa Monica, California, January 1963.
42. E.H. Benstein and W.Q. Wagner, "Exercising Propulsion DTLCC Methods in Advanced Development Engines", Teledyne Report No.

1558; Proceedings of the Department of Defense Symposium on Engine Life Cycle Cost; Warminster, PA, U.S.A., May 1978.
43. A.H. Berard, "Life Cost Management, Methodology, and Case Studies", Defense Systems Management School, Fort Belvoir, VA, U.S.A., 1977, Report No. AD-A052388, 77/10/100, 78 N24999. Available from National Technical Information Service, Springfield, VA 22161, USA.
44. R.E. Biedenbender, "Cost-to-Produce", Proceedings of the Annual Reliability and Maintainability Symposium, 1973, pp. 481–483.
45. J.E. Biegel, "A Multivariate Modularization Technique Designed to Minimize Life Cycle Costs for Large Systems", Proceedings of the 1979 Spring Annual Conference, American Institute of Industrial Engineers, Inc., San Francisco, California, May 22–25, 1979, pp. 335–342.
46. J.E. Biegel, and B. Bulcha, "System Modularization to Minimize Life Cycle Costs", Report No. RADC-TR-77-13, 1977, Rome Air Development Center, Griffis Airforce Base, NY 13441, USA.
47. J.E. Biegel, and B. Bulcha, "Multi level Modularization of Systems to Minimize Life Cycle Cost", Report No. RADC-TR-78-207, 1978, Rome Air Development Center, Griffis Airforce Base, NY 13441, U.S.A.
48. B.S. Blanchard, "Life Cycle Costing: A Management Concept For System/Product Acquisition", A Report from Virginia Polytechnic Institute and State University, Blacksburg, Virginia 24061, U.S.A., 1978.
49. B.S. Blanchard, "Life Cycle Costing—A Review", Terotechnica, Vol. 1, 1979, pp. 9–15.
50. B.S. Blanchard, "Cost Effectiveness, System Effectiveness, Integrated Logistics Support, and Maintainability", IEEE Transactions on Reliability, Vol. R-16, 1967, pp. 117–125.
51. B.S. Blanchard, "Design and Manage to Life Cycle Cost", M/A Press, Forest Grove, Oregon, USA, 1978.
52. B.S. Blanchard, "Cost Effectiveness Analysis Concept/Applications/ Rish and Uncertainty/Case Study Approach", A Report from Virginia Polytechnic Institute and State University, Blacksburg, Virginia 24061, USA, 1978.
53. H.S. Blanks, "Traffic Signal Equipment Procurement Based on Cost of Ownership and Community Cost", Quality and Reliability Engineering International, Vol. 1, 1985, pp. 195–207.
54. G.W. Bleasdale, "Some Engineering Aspects of Life Cycle Costing", AGARD Conference Proceedings No. 289 (Design to Cost and Life Cycle Cost), Paper Presented at the Flight Mechanics Panel Symposium, Amsterdam, The Netherlands, May 1980, pp. 1–17.
55. S.J. Blewitt, "Cost and Operational Effectiveness of R&M Improvements", Proceedings of the Annual Reliability and Maintainability Symposium, 1978, pp. 417–421.
56. R.D. Bliesner, J. Keller, G.Z. Watters and B.W. Cone, "Economic Analysis of Life Cycle Costing Irrigation Pipe Network Design", ASAE

National Energy Symposium, Vol. 2, Kansas City, USA, September 29–October 1, 1980, Published by ASAE, St. Joseph, Michigan, USA, 1981, pp. 456–460.

57. A. Blumstein, "The Choice of Analytical Techniques in Cost-Effectiveness Analysis", Institute for Defense Analyses, Washington, D.C., USA, October 1965.
58. M.B. Blunn, "Function Cost Analysis", First International Congress on Terotechnology, London, U.K., May 1979, pp. 1–2.
59. W. H. Boden, "Designing for Life Cycle Cost", Defense Management Journal, Vol. 12, No. 1, January 1976, pp. 29–37.
60. B.W. Boehm, "Software Life Cycle Factors", in Handbook of Software Engineering, Edited by C. R. Vick and C.V. Ramamoorthy, Van Nostrand Reinhold Company, New York, 1984, pp. 494–518.
61. Boeing Aircraft Company, "Life-Cycle Cost/System Effectiveness Evaluation and Criteria", Document No. D180-17648-1, Boeing Aircraft Company, Seattle, USA, 1974.
62. A. Boerckel, "The Formula for Survival—Optimum Quality at Optimum Cost", Proceedings of the Annual Reliability and Maintainability Symposium, 1973, pp. 373–378.
63. L. Borghagen and L. Brinkhagen, "LCC-Procurement at the Swedish State Railways", Proceedings of the Annual Reliability and Maintainability Symposium, San Francisco, 1984, pp. 349–358. Available from IEEE, New York, U.S.A.
64. B. Bowen, J. Williams, "Life-Cycle Costing and the Problems of Data", Industrialization Forum, Vol. 6, No. 3–4, 1975, pp. 21–24.
65. T.N. Bowers, "Reliability Guarantee Versus Cost of Capital", Proceedings of the Tenth Annual West Coast Reliability Symposium, Century City, California, February 21, 1969, pp. 157–164. Available from Western Periodicals Company, 13000 Raymer Street, North Hollywood, California 91605, U.S.A.
66. B.D. Bradley, "Incentives for Cost Efficiency in the Design and Acquisition of Avionic Systems", Report No. WN-8239-ARPA, The Rand Corporation, Santa Monica, California, May 1973.
67. R.C. Brannon, "Army Life Cycle Cost Model", NTIS Report No. AD-A021-900, January 1976. Available from the National Technical Information Service (NTIS), Springfield, Virginia, USA.
68. R.C. Brannon, "Life Cycle Costing Procurement Guide LCC1, LCC2, LCC3", U.S. Department of Defense, Washington, D.C., July 1970.
69. W.G. Briggs, J.B. DeWolf, and R.T. Lund, "Life-Cycle Costs: Concepts, Considerations and Cases", Center for Policy Alternatives, Massachusetts Institute of Technology (MIT), Massachusetts, November 1975, USA.
70. W.R. Briggs, "SWECS Cost of Energy Based on Life Cycle Costing", Atomic International Division, Golden Company, Rocky Flats Plant,

May 1980, 55 pp, NTIS No. RFP-3120/3533. Available from National Technical Information Service, Springfield, Virginia.

71. H.A. Brode, "Cost of Ownership—An overview—Life Cycle Costs: Evaluation of Avionic System Reliability Improvements", Proceedings of Annual Reliability and Maintainability Symposium, 1975, pp. 212–216.
72. C. Brook, R. Barasia, "A Support System Life Cycle Cost Model", Proceedings of the Annual Reliability and Maintainability Symposium, 1977, pp. 297–302. Available from IEEE, New York, U.S.A.
73. A.M. Brown, "Present Worth of Future Energy Conservation—A Discussion on Economic Life of Equipment", Australian Refrigeration, Air Conditioning and Heating, Vol. 31, No. 3, March 1977, pp. 7–15.
74. F.D. Brown, "Life Cycle Cost of C-130E Weapon System", Report No. 77-46, Brooks Air Force Base, Texas, 1975.
75. R.J. Brown, "A New Marketing Tool: Life-Cycle Costing", Industrial Marketing Management, Vol. 8, 1979, pp. 109–113.
76. R.J. Brown and R.R. Yanuck, "Life Cycle Costing: A Practical Guide for Energy Managers", The Fairmont Press, Atlanta, Georgia, USA, 1980, 299 pp.
77. S.R. Bruins, P. Timberman, "Solar System Cost/Performance Analysis", Proceedings of the Annual Meeting, American Section of the International Solar Energy Society, Vol. 1, Orlando, Florida, USA, June 6–19, 1977, Section 8, pp. 31–34.
78. E.R. Brussell, K. Pope and K.M. Tasugi, "Cost of Ownership—Industry Viewpoint—Parametric Analysis of Operating and Support", Proceedings of the Annual Reliability and Maintainability Symposium, 1975, pp. 217–220.
79. N.S. Bryan, "Contracting for Life Cycle Cost to Improve System Affordability", Proceedings of the Annual Reliability and Maintainability Symposium, 1981, pp. 342–345.
80. O. Bryk, "Models in Cost-Effectiveness Analysis: An Example", RAC Paper P-2, Research Analysis Corporation, McLean, Virginia, 1965.
81. M.H. Burmeister, "Parametric Analysis of Design to Life Cycle Cost", IEEE Proceedings of the National Aerospace and Electronics Conference, NAECON 1980, Vol. 2, Dayton, Ohio, May 1980, pp. 675–681.
82. Canadian Forces Procedure CFP 113, "Life Cycle Management System-Guidance Manual", Department of National Defence, Ottawa, Canada, 1 September, 1976.
83. F.R. Carey, "System Reliability Cost Tradeoff Methodology", Proceedings of the Tenth Annual West Coast Reliability Symposium, Century City, California, February 21, 1969, pp. 91–104. Available from Western Periodicals Company, 13000 Raymer Street, North Hollywood, California 91605, U.S.A.
84. F.T. Carlson, "Design to Life Cycle Cost Research", in Design to Cost and Life Cycle Cost, AGARD Conference Proceedings No. 289, May

1980, pp. 5-1 to 5-15. Available from National Technical Information Service (NTIS), 5285 Port Royal Road, Springfield, Virginia 22161, U.S.A.

85. P.R. Carpenter, "A Preliminary Analysis of Energy-Efficient Procurement Standards and Life-Cycle Costing", A Report for the New England Energy Management Information System at Massachusetts Institute of Technology (MIT), Technical Report No. MIT-NEEMIS-77001 TR, Cambridge, Massachusetts, USA, January 1977.
86. T.V. Carver, "New Maintenance Manuals Help to Reduce Life Cycle Costs", Defense Management Journal, Vol. 14, No. 5, 1978, pp. 26–31.
87. T.V. Carver, "Life Cycle Cost: Attitude and Latitudes", Defense Management Journal, Vol. 15, No. 4, 1979, pp. 12–17.
88. Center for Policy Alternatives, Massachusetts Institute of Technology (MIT), Massachusetts," Consumer Use of Life-Cycle Cost Information", A Report Prepared for the National Science Foundation, USA, August, 1977.
89. F.M. Chakour, "Automated Life Cycle Cost Models for Army Weapon Systems", US Army Weapons Command, Rock Island, Illinois, May 1969.
90. N.E. Chang, "Life Cycle Cost Evaluation of Underground Distribution Systems", Specifying Engineering, Vol. 42, No. 5, October 1979, pp. 1102–1104.
91. R.P. Charette, "Life Cycle Cost Analysis for Buildings", ASHRAE Transactions, 1980, Part 2, pp. 167–179.
92. C.F. Chavez, "Diagnostic Programming: Costs and Benefits", Proceedings of the Annual Reliability and Maintainability Symposium, 1982, pp. 25–29.
93. R.C. Chawla, M.M. Varma, "Data Sources, Reliability, Updating-Inflation and Relevant Indices", American Institute of Chemical Engineers, Modular Instruction Series G: Design of Equipment, Vol, 1: Plant Design and Cost Estimating, Series Editor: James Beckman, 1986, pp. 72–83.
94. R. Chisholm, "Life Cycle Cost Analysis (LCCA) in Military Aircraft Procurement", in Design to Cost and Life Cycle Cost, AGARD Conference Proceedings No. 289, May 1980, pp. 1-1 to 1-8. Available from National Technical Information Service (NTIS), 5285 Port Royal Road, Springfield, Virginia 22161, U.S.A.
95. A. Chop, "Capability Measures for System Effectiveness", Final Technical Report, RADC-TR-72-26, Prepared Under Contract F30602-70-C-0197 by Lockheed Missiles & Space Company, Sunnyvale, California, February 1972.
96. J.J. Clark, R.H. Penner, "Life Cycle Costing in the Hotel Industry", Industrialization Forum, Vol. 6, No. 3–4, 1975, pp. 25–32.
97. L.P. Clark, "A Life-Cycle Cost Analysis Methodology for Fire Protection Systems in New Health Care Facilities", Report No. NBSIR

82-2558, July 1982, National Bureau of Standards, Department of Commerce, Washington, D.C. 20234. Available from National Technical Information Service (NTIS), Springfield, VA 22161, USA.

98. A.S. Clorfeine, "Economic Feasibility of Photovoltaic Energy System", Conference Record, 14th IEEE Photovoltaic Spec. Conf., San Diego, California, January 1980, pp. 986–989.
99. W.J. Coad, "Investment Optimization, A Methodology for Life Cycle Cost Analysis", ASHRAE Journal, Vol. 19, January 1977, pp. 29–33.
100. W.J. Coad, "Life Cycle Costing", Heating/Piping/Air Conditioning, Vol. 47, No. 13, December 1975, p. 62.
101. C.K. Coe, "Life Cycle Costing by State Governments", Public Administration Review, Vol. 41, September–October, 1981, pp. 564–569.
102. G.B. Cohen, "Reliability Influence on Life Cycle Costs", Annals of Assurance Sciences, (Reliability and Maintainability Conference), 1969, pp. 452–459. Distributed by Gordon and Breach Science Publishers, New York, U.S.A.
103. J. Cohen, "Life Cycle Performance of Clothes Dryers", Proceedings of the Annual Reliability and Maintainability Symposium, 1980, pp. 91–94.
104. C.A. Collier and D.E. Jacques, "Optimum Equipment Life by Minimum Life-Cycle Costs", Journal of Constructional Engineering Management, Vol. 110, No. 2, June 1984, pp. 248–265.
105. D.E. Collins, "Analysis of Available Life Cycle Cost Models and Their Applications", Report by AFSC/AFLC Commanders Working Group, Wright Patterson Air Force Base, Ohio, USA, 1976.
106. Collins Radio Company, "Life Cycle Costs", Research Study by Collins Radio Company, Cedar Rapids, Iowa, August 1966.
107. D.E. Collins, "Models a Key to Air Force Life Cycle Cost Implementation", Defense Management Journal, Vol. 12, No. 1, January 1976, pp. 54–59.
108. D.H. Comey, D.G. Culy, and J.E. Cassidy, "Tradeoff Studies with an Interactive Engine/Airframe Life-cycle-cost Model", Society of Automotive Engineers (SAE), Aerospace Meeting, San Diego, California, U.S.A., November 27–30, 1978, 10 pages, SAE Paper 781033.
109. Commonwealth of Australia, "Life Cycle Costing and Energy Conservation", Commonwealth of Australia, Department of Transport and Construction, Australia, 1982, 34 pp.
110. W.S. Copes, "Conversion of MICOM, Time-Phased Life Cycle Cost Estimating Model from Cobal to Fortran IV", Report No. AMSAA-TM-131, Army Material Systems Analysis Agency, Washington, D.C., U.S.A., March 1972.
111. A. Coppola, "Reliability as a Capital Investment", Proceedings of the Annual Reliability and Maintainability Symposium, 1974, pp. 349–357. Available from IEEE, New York, U.S.A.

112. "Cost Engineering in the Process Industries", Edited by Cecil H. Chilton and the Staff of Chemical Engineering, McGraw-Hill Book Company, Inc., New York, USA, 1960.
113. J.R. Coult, et al., "Aircraft Avionics Tradeoff Study", Volume II: Concept Development and Tradeoff, Part 2, Equipment Tradeoffs, Honeywell, Inc., Systems and Research Center, Minneapolis, Minnesota, USA, September 1973.
114. W.D. Cowie, E.E. Abell, E.W. Horn, "A New Development Concept for Gas Turbine Engine Optimize Life Cycle Costs", Proceedings of the International Symposium on Air Breathing Engines, Munich, West Germany, March 7–12, 1976, pp. 27–38.
115. T.D. Cox, B.L. Morrison, "An Approach for Determining Optimum RAM Requirements", Proceedings of the Annual Reliability and Maintainability Symposium, 1977, pp. 378–384.
116. R.K. Crowe, P.S. Kilpatrick and W.C. Marshall, "Aircraft Avionics Tradeoff Study," Volume III: Concept Application, Evaluation, and Implementation, Honeywell Inc., Systems and Research Center, Minneapolis, Minnesota, September 1973.
117. C.E. Curry, R.V. Earle, G.H. Pedersen, "Turbine Engine Cost Reduction Using Life Cycle Cost Techniques", Society of Automotive Engineers, Aerospace Meeting, San Diego, California, November 27–30, 1978, 9 pages. SAE Paper No. 781031.
118. C.E. Curry, "Life Cycle Cost as a Propulsion System Design Consideration", American Society of Mechanical Engineers, Gas Turbine Conference and Products Show, Philadelphia, PA, USA, March 27–31, 1977, 7 pages. ASME Paper 77-GT-99.
119. C.E. Curry, "Engine Life Cycle Cost Modeling in the Conceptual Phase", 11th Propulsion Conference, American Institute of Aeronautics and Astronautics and Society of Automotive Engineers, Anaheim, California, September 29–October 1, 1975, 8 pages. AIAA Paper 75-1288, 75/09/00, 75N 5678.
120. R. Cyert, "Managing to Get Into the Details", Mechanical Engineering, Vol. 109, No. 1, January 1987, pp. 56–64.
121. A.F. Czajkowski, "Managerial Decision-Making in Establishing R&M Design Goals", IEEE Transactions on Reliability, Vol. R-32, No. 3, August 1983, pp. 253–258.
122. T.A. Daly and P.H. Ockerman, "Commercial Reliability Programs—A Good Investment", Proceedings of the 11th National Symposium on Reliability and Quality Control, 1965, pp. 385–391.
123. A.M. Dame, "Determining Total Equipment Cost", Proceedings of the Second National Convention on Military Electronics, Washington, D.C., July 16–18, 1958.
124. M.B. Darch, "Selling Reliability Engineering to the Company", Microelectronics and Reliability, Vol. 19, No. 1, 1979, pp. 1–5.

125. G.H. Davidson, R.E. Griffiths, "An Identification and Characterization of Cost Models/Techniques Used by Air Force Logistics Command to Estimate Jet Engine Operation and Support Costs", School of Systems and Logistics, Air Force Institute of Technology, Wright-Patterson AFB, Ohio, USA, 1977, AD-AO 440833 AFIT-LSSR-01-77A, 77/06/00, 78 N12095. Available from National Technical Information Service (NTIS), Springfield, Virginia, USA.
126. W.L. Davidson, B.J. Landstra, "A Logistics Analysis and Ranking Model ALARM, Proceedings of the Annual Reliability and Maintainability symposium, 1975, pp. 1–14.
127. A. Davies, K.J. Skinner, "The Application of a Life Cycle Cost Model to Modular Electronic Systems", The Radio and Electronic Engineer, Vol. 53, No. 5, May 1983, pp. 209–215.
128. J.G. Davies, "Life Cycle Costing for Plant Construction Projects", Planning Engineering, Vol. 32, No. 22, 1978, pp. 121–124.
129. P.H. Day, "Capital Investment Appraisal for Mechanical and Electrical Services in Commercial Buildings", The Heating and Ventilating Engineer, February 1976, pp. 4–7.
130. L. DeAngelis, R. Bell, "Life Cycle Reliability Programs for Rocket Motors of US Army Missile System", US Army Material Command, Washington, D.C., USA, May 1972.
131. D.E. DeBurkarte, "A Standard Life Cycle Cost Model for Inertial Systems", Proceedings of the IEEE National Aerospace and Electronics Conference (NAECON'76), Dayton Convention Center, May 18–20, 1976, pp. 687–695.
132. R.W. Decicco, "Economic Evaluation of Research Projects—By Computer", in Modern Cost Engineering Techniques, Edited by Herbert Popper and the Staff of Chemical Engineering, McGraw Hill Book Company, New York, USA, 1970, pp. 322–328.
133. L.O. Degelman, "Computer Optimization of Solar Collector Area Based on Life-cycle Costing", Proceedings of the Annual Meeting American Section of the International Solar Energy Society, Vol. 1, Orlando, Florida, USA, June 6–19, 1977, Section-8, Session J.2, 5 pp.
134. Department of Industry, "Management Aspect of Terotechnology-Life Cycle Costing", London, U.K., 1975.
135. Department of National Defense, "Canadian Forces Life Cycle Cost Casebook", Report No. MMI 1106, June 30, 1984, Department of National Defense, Ottawa, Ontario, Canada.
136. Design Practice Committee of the Illuminating Engineering Society, "Life Cycle Cost Analysis of Electric Lighting Systems", Lighting Design and Application (IES Transaction), May 1980, pp. 43–48.
137. J.J. Deringer, "Energy Conservation Measures for Commercial Buildings Used in Life Cycle Cost Analysis", Proceedings of the 15th Inter

Society Energy Conversion Engineering Conference, Seattle, Vol. 3, August 1980, pp. 2350–2355.

138. J.H. Dever, "Evaluation of Disposable Hydraulic Filter Elements for F/RF-101 Life Cycle Cost Procurement", Naval Air Development Center, Warminster, Pennsylvania, January 1973.

139. W.G. Devlin, "Strategic Bomber Weapon System Cost Model", Report No. D6-19516, The Boeing Aerospace Company, Seattle, USA, February 1968.

140. W.G. Devlin, "Tactical Fighter Weapon System Costing/Initial Investment and Annual Operations and Maintenance", Report No. D6-9909, The Boeing Aerospace Company, Seattle, USA, March 1964.

141. B.S. Dhillon, "Life Cycle Cost: A Survey", Microelectronics and Reliability, Vol. 21, No. 4, 1981, pp. 495–511.

142. B.S. Dhillon, Reliability Engineering in Systems Design and Operation, Van Nostrand Reinhold Company, New York, 1982.

143. B.S. Dhillon, H. Reiche, Reliability and Maintainability Management, Van Nostrand Reinhold Company, New York, 1985.

144. D.B. Dickinson, L. Sessen, "Life Cycle Cost Procurement of Systems and Spares", Proceedings of the Annual Reliability and Maintainability Symposium, 1976, pp. 282–286. Available from IEEE, New York, U.S.A.

145. W.H. Dickinson, "Economic Evaluation of Industrial Power Systems Reliability", Trans. of AIEE on Applications and Industry, Vol. 76, Part II, November 1957, pp. 264–272.

146. J.F. Digby, "DOD Actions to Control Avionics Life Cycle Costs", Report No. WN-8234-ARPA, The Rand Corporation, Santa Monica, California, USA, May 1973.

147. R.D. Dighton, "The Hornet Program: A Design to Life Cycle Cost Case Study", in Design to Cost and Life Cycle Cost, AGARD Conference Porceedings No. 289, May 1980, pp. 8-1 to 8-12. Available from National Technical Information Service (NTIS), 5285 Port Royal Road, Springfield, Virginia 22161, U.S.A.

148. D.M. Dix, F.R. Riddell, "Projecting Cost-Performance Trade-offs, for Military Vehicles", Astronautics and Aeronautics Journal, Vol. 14, No. 9, Sept. 1976, pp. 40–50.

149. J.G. Dixon, R.H. Anderson, "Implementation of the Design-to-Cost Concept", Proceedings of the Annual Reliability and Maintainability Symposium, 1975, pp. 224–229.

150. R.D. Doering, R.D. Evans, Y.A. Hosni, "Saving Tax Dollars in Public Buildings—Case Studies in Florida", Fall Industrial Engineering Conference AIIE Proceedings, Houston, Texas, November 13–16, 1979. Published by AIIE (AIIE-P-359), Norcross, Ga, 1979, pp. 55–59.

151. B.A. Donahue, D.W. Sarver, E.M. Bellino, "Field Test of a Life-Cycle

Cost Analysis Method for Solvent Management", Technical Report, US Army Corps Engineering, Construction Engineering Research Laboratories, N-86-21, September 1986, 18 pp.

152. J.J. Dougherty, G.W. Hogg, "Army Helicopter Mission Reliability and Cost Analysis", Proceedings of the Annual Reliability and Maintainability Symposium, 1983, pp. 411–416.

153. L.E. Dover, B.E. Oswald, Jr., "A Summary and Analysis of Selected Life Cycle Costing Techniques and Models", Masters Thesis, Air Force Institute of Technology, Wright-Patterson Air Force Base, Ohio, U.S.A., 1974.

154. W.C. Dretz, "Preliminary Design Aspects of Design-to-cost for the YF-16 Prototype Fighter", in AGARD CP-147, Aircraft Design Integration and Optimization, USA, October, 1973/June 1974.

155. D.J. Dreyfuss, "A Survey of Costing Methods in the Avionics Industry", Report No. WN-8235-ARPA, The Rand Corporation, Santa Monica, California, USA, May 1973.

156. S. Duham, "A Cost Effective Approach to Integrated Logistics Support", Logistics Spectrum, Vol. 5, No. 3, Fall 1971, pp. 12–17.

157. S. Duham, J. Catlin, "Total Life Cost and the 'Ilities'", Proceedings of the Annual Reliability and Maintainability Symposium, 1973, pp. 491–495.

158. D.P. Dunbar, "Reliability of System Design and Testing", Defense Management Journal, Vol. 14, No. 6, 1978, pp. 26–30.

159. E.P. Durbin, "Designing to Low Ownership Cost Requires Knowledge of Many Factors", Defense Management Journal, Vol. 9, No. 3, July 1973, pp. 21–24.

160. A. Dushman, "Influence of Reliability on Life Cycle Costs", American Society of Mechanical Engineers, Paper No. 70-DE-43, January 1970. Available from the American Society of Mechanical Engineers, New York.

161. A. Dushman, "Review of the Application of Life Cycle Costing to the ARC-XXX/ARC-164", Report Avionic Program Office, Wright Patterson Air Force Base, Ohio, 1975.

162. A. Dushman, "Life Cycle Costing", Report No. 533–0758985, Collins Radio Co., Colorado, 10th August, 1966.

163. D.R. Earles, "LCC-Commercial Application, Ten Years of Life Cycle Costing", Proceedings of the Annual Reliability and Maintainability Symposium, January 1975, pp. 74–85. Available from IEEE, New York, U.S.A.

164. D.R. Earles, "Balancing Life Cycle Acquisition and Support Costs—A User/Producer Relationship", Proceedings of the Annual Society of Logistics Engineers Symposium, September 1974, pp. 110–115.

165. D.R. Earles, "Techniques for a Multi faceted Discipline", Defense Management Journal, 1976, pp. 38–47.

166. D.R. Earles, "Design to Operation and Support Costs", Proceedings of the Annual Symposium on Reliability and Maintainability, Los Angeles, California, USA, 29–31 January 1974, pp. 149–153.
167. D.R. Earles, "Integrated Logistic Support and Life Cycle Costing", Report No. T 437, Missile Systems Division, Raytheon Company, Huntsville, Alabama, May 1969.
168. M.E. Earles, "Factors, Formulas and Structures for Life Cycle Costing", (Second Edition), Eddins-Earles, Concord, Mass., 1981.
169. H. Ebenfelt, S. Orgen, "Some Experiences from the Use of an LCC Approach", Proceedings of the Annual Reliability and Maintainability Symposium, 1974, pp. 142–146.
170. H. Ebenfelt, O. Waak, "LCC-Defense Application, Some Comments on LCC as a Decision Making Tool", Proceedings of the Annual Reliability and Maintainability Symposium, 1975, pp. 63–73.
171. P.S. Eisenhut, "New Insights into the Life Cycle Approach", AIIE Transactions, Vol. 5, No. 2, June 1973, pp. 150–155.
172. L.F. Eliel, "Logistics Effect Model (LEM) Applications", Proceedings of the Annual Reliability and Maintainability Symposium, 1978, pp. 260–266.
173. J.M. English, "Cost-Effectiveness—The Economic Evaluation of Engineering Systems", John Wiley & Sons, New York, USA, 1968.
174. W. J. Fabrycky, "Designing for the Life Cycle", Mechanical Engineering, Volume 109, No. 1, January 1987, pp. 72–74.
175. D.R. Fairbanks, "A Computer Program to Determine the Optimum Configuration of Solar Assisted Building Heating and Cooling Systems Based Upon Life-cycle Cost", Presented at Intersociety Energy Conversion Engineering Conference, August 18–22, 1975, p. 202.
176. Federal Trade Commission, "Trade Regulation Rule, Relating to Incandescent Lamps (Light Bulbs)—Disclosure of Lumens, Life, Cost and Other Data", Federal Trade Commission, Washington, D.C., January 1971.
177. A.M. Feiler, "Incorporating Risk Analysis into Life Cycle Costing", Proceedings of the Annual Reliability and Maintainability Symposium, 1986, pp. 469–476.
178. W.F. Finian, "LCC Arrives for Major Systems", Defense Management Journal, Vol. 5, No. 1, Winter 1968–1969, pp. 28–35.
179. M. Fiorello, "Getting Real Data for Life Cycle Costing", Report No. P-5345, The Rand Corporation, Santa Monica, California, January 1975.
180. M. Fiorello, "Estimating Life-cycle-costs—A Case Study of the ATD", (NTIS), Report No. R-1518-PR, US Department of Commerce, February 1975. Available from the National Technical Information Service (NTIS), Springfield, Virginia.
181. M. Fiorello, "Data System for Life Cost Information: Some Issues and

Recommendations", Report No. WN-8238-ARPA, The Rand Corporation, Santa Monica, California, May 1973.
182. G.H. Fisher, Cost Considerations in Systems Analysis, American Elsevier Publishing Company, New York, 1971.
183. G.H. Fisher, and H.J. Slovin, "Life Cycle Cost Impact on High Reliability Systems", Proceedings of the Annual Reliability and Maintainability Symposium, 1974, pp. 358–362.
184. C.M. Fishman, H.J. Slovin, "Life Cycle Cost Impact on High Reliability Systems", Proceedings of the Annual Reliability and Maintainability Symposium, 1974, pp. 358–362. Available from IEEE, New York, U.S.A.
185. T.J. Fody, "The Procurement of Window Air Conditioners Using Life Cycle Costing", Proceedings of the Annual Reliability and Maintainability Symposium, 1977, pp. 81–88. Available from IEEE, New York, U.S.A.
186. J.B. Forsythe, "Light Rail/Rapid Transit: New Approach for the Evaluation of Energy Savings—1, Life Cycle Cost from Synthetic Routes/Operational Models", IEEE Trans. on Ind. Appl., Vol IA-16, No. 5, September/October 1980, pp. 655–665.
187. B. Fox, "Total Annual Cost, A Reliability Criterion", Proceedings of the 10th National Symposium on Reliability and Quality Control, Washington, D.C., 1964, pp. 119–124.
188. K.A. Frederiksen, "Incentives for Reliability", Tenth National Symposium on Reliability and Quality Control, Washington, D.C., January 1964, pp. 210–213.
189. S. French, "Decision Analysis and Life Cycle Costing", in Electronic Systems Effectiveness and Life Cycle Costing, Edited by J.K. Skwirzynski, Springer-Verlag, Berlin, 1983, pp. 633–646.
190. D.V. Fricker, "Life Cycle Cost: The Key Roles of Reliability and Maintainability", Engineering, Vol. 219, No. 2, February 1977, pp. 164–166.
191. B. Frumkin, P.W. Hodge, M.J. Sabia, "An Econometric Approach to Redundancy Selection", Grumman Aerospace Corporation, Beth Page, N.Y., U.S.A., March 1972, pp. 1–8. Report No. 347.40.00.00-K4-27.
192. C. Fullman, "Accuracy is the Best Policy for Life Cycle Analysis", Specifying Engineering, Vol. 41, No. 3, 1979, pp. 127–129.
193. I.J. Gabelman, "Methodology for Control of Life Cycle Costs for Avionics Systems", in Design to Cost and Life Cycle Cost, AGARD Conference Proceedings No. 289, May 1980, pp. 23-1 to 23-8. Available from National Technical Information Service (NTIS), Springfield, Virginia, U.S.A.
194. M.P. Galin, C. Marecki, and A.E. Schutzman, "Life Cycle Cost/Design-to-Cost Planning, Applications, and Methods", Report No. MTR-3032, The MITRE Corp., Bedford, MA, U.S.A., June 1975.

195. J. Gallois, "Medium Speed Diesel Engines—Total Energy Calculations and Life Cycle Costing", Trans. Inst. Mar. Eng., Vol. 93, 1981, pp. 1–15.
196. V. Ganapathy, "Life Cycle Costing Applied to Motor Selection", Process Engineering, London, Vol. 64, No. 7, July 1983, pp. 51–52.
197. P.E. Gannon, "Costs of Interruptions, Economic Evaluation of Reliability", Proceedings of the 1977 Reliability Conference for the Electric Power Industry, 1977, pp. 60–65.
198. P.E. Gannon, "Supplemental Cost Data: Examples of Reliability And Availability Analysis of Common Low-Voltage Industrial Power Distribution Systems", Proceedings of the 1977 Reliability Conference for the Electric Power Industry, 1977, pp. 81–85.
199. J.S. Gansler, "Application of Life Cycle Costing to the DOD System Acquisition Decision Process", Proceedings of the Annual Reliability and Maintainability Symposium, 1974, pp. 147–148.
200. J.S. Gansler, "Acquisition Objective Changes from One of Sophistication to Reliability at Lower Cost", Defense Management Journal, Vol. 9, No. 3, July 1973, pp. 3–6.
201. D. Garribba, C. Joppolo and G. Volta, "Energy and Economic Cost of a Washing Machine: Statistical Life Cycle Analysis for Energy Saving", International Conference on Energy Use Management, Los Angeles, California, October 1979.
202. C.E. Garrison, "Technological Progress in Electronic Components: A Life Cycle Support Problem", Defense Systems Management Review, Vol. 1, No. 1, 1976, pp. 25–38.
203. General Dynamics Corporation, "Aircraft Avionics Environmental Control Design and Analysis Procedures for Optimized Life Cycle Cost", Report. No. GDCA-DDB-71-003, Design Manual, General Dynamics Corporation, USA, July 1971.
204. General Services Administration, "Life Cycle Costing—Work Book", Federal Supply Service, Washington D.C., February 1977.
205. R.M. Genet, "Generalized Conclusions Concerning Life Cycle Costs of Air Force Inertial Navigation Systems", Report, Aerospace Guidance and Metrology Center, Newark Air Force Station, Newark, Ohio, U.S.A., October 1972.
206. R.M. Genet, "Procurement and Maintenance of Inertial Instruments and Systems in the Royal Swedish Air Force", Report Aerospace Guidance and Metrology Center, Newark Air Force Station, Newark, Ohio, U.S.A., October 1972.
207. R.M. Genet, T.D. Meitzler, "Avionics Proliferation, A Life Cycle Cost Perspective", Defense Management Journal, Vol. 12, No. 1, January 1976, pp. 60–64.
208. R.M. Genet, "Avionics Cost Reduction Through Improved Tests", Proceedings of the IEEE 1974 National Aerospace and Electronics

Conference, NAECON' 74, Dayton Convention Center, May 13–15, 1974, pp. 534–538.

209. J.L. George and F.J. Milner, "Total Life Cycle Cost of Ownership Study for AN/URC-78 (XE-1)/V", Report, US Army Electronics Command, Fort Manmouth, New Jersey, 1972.
210. J.L. George and F.J. Milner, "Total Life Cycle Cost of Ownership Study for AN/URC-78(XE-1)/V", Report, US Army Electronics Command, Fort Manmouth, New Jersey, 1972.
211. J.D.S. Gibson, D.E. Collins, C.A. Conley, "Engine Life Cycle Cost Considerations During the Validation Phase", 11th Propulsion Conference, American Institute of Aeronautics And Astronautics and Society of Automotive Engineers, Anaheim, California, U.S.A., September 29–October 1, 1975, 11 pages. AIAA Paper 75-1289, 75/09/00, 76A10279.
212. A.S. Goldman, "Problem in Life Cycle Support Cost Estimation", Naval Research Logistics Quarterly, Vol. 16, 1969, pp. 111–120.
213. A.S. Goldman, T.B. Slattery, "Maintainability: A Major Element of System Effectiveness", John Wiley & Sons, Inc., New York, 1964.
214. M.B. Goldman, R.W. Tipton, "Accurate LCC Estimating Early in Program Development", Proceedings of the Annual Reliability and Maintainability Symposium, 1974, pp. 256–265. Available from IEEE, New York, U.S.A.
215. T.A. Goldman (Ed.), Cost-Effectiveness Analysis, Praeger, New York, 1971.
216. B.S. Goldsberg, A.G. Hausser, B.T. Le, "Accelerated Cycle Life Testing of Lead-Acid Golf Car Batteries and the Influence of Separator Type on Battery Life, Energy Consumption and Operation Operating Cost", Journal of Power Sources, Vol. 10, No. 2, July 1983, pp. 137–148.
217. M. Goldstein, "Economics of Protected AC Power for Computers", Proceedings (INTELEC) of International Telecommunications Energy Conference, Washington, DC, November 1979, pp. 122–126.
218. D.J. Gordon, "A Simplified Approach to Life Cycle Costing for Plant Construction Projects", Plant Engineering, vol. 32, 1978, pp. 121–126.
219. K.K. Govil, "New Analytical Models for Logistics Support and Life Cycle Cost Versus Reliability Function", Microelectronics and Reliability, Vol. 24, No. 1, 1984, pp. 61–63.
220. J.S. Greenberg, "The Economic Implications of Unreliability", Proceedings of the Annual Reliability and Maintainability Symposium, 1976, pp. 293–299.
221. G.R. Greenfield, "Cost Analysis of Reliability Programs", Proceedings of the Seventh Annual West Coast Reliability Symposium, May 14, 1966, pp. 93–107. Available from Western Periodicals Company, 13000 Raymer Street, North Hollywood, California, USA.
222. H. Grieser, "Impact on System Design of Cost Analysis of Specifications and Requirements", in Design to Cost and Life Cycle Cost,

AGARD Conference Proceedings No. 289, May 1980, pp. 6-1 to 6-10. Available from National Technical Information Service (NTIS), 5285 Port Royal Road, Springfield, Virginia 22161, U.S.A.

223. J.W. Griffith, B.J. Keely, "Life Cycle Cost Benefit Analysis", Report No. PB251848, 1976, Available from National Technical Information Service (NTIS), Springfield, Virginia, USA, 1976.
224. J.W. Griffith, "Benefits of Daylighting—Cost and Energy Savings," ASHRAE Journal, Vol. 20, No. 1, January 1978, pp. 53–56.
225. J.W. Griffith, "Daylight Utilization in Passive Solar Systems Using Life-Cycle Cost-Benefit Analysis", pp. 316–322, K.G. Associates, U.S.A., NTIS No. LA-6637-C.
226. J.W. Griffith, "Resource Optimization Calls for Analysis Based on Life-Cycle Costs", Professional Engineering, Vol. 45, No. 6, June 1975, pp. 39–41.
227. J.W. Griffith, Life Cycle Cost Workbook, Report No. NBS-GCR-79-186, December 1979. Available from National Technical Information Service (NTIS), Springfield, Virginia, VA 22161, USA.
228. R.W. Grimm, "Fire Control Radar and Airborne Computer Cost Prediction Based on Technical Parameters", Proceedings of the IEEE 1974 National Aerospace and Electronics Conference (NAECON'74), held at Convention Center, Dayton, Ohio, 13–15 May 1976, pp. 506–513.
229. J.F. Grosson, "Designing to a Cost: An Update on Navy Initiatives", Defense Management Journal, Vol. 15, No. 1, 1979, pp. 7–10.
230. M.S. Grover, M.T.G. Gillespie, "A Cost Oriented Reliability and Maintainability Assessment of Circuit Breakers in the Design of Auxiliary Electrical Power Systems of a Generating Station", Proceedings of the Third Annual Reliability Engineering Conference for the Electric Power Industry, 1976, pp. 90–95.
231. C.F. Gullo, L.N. Whited, "Evaluating Indirect Costs Through the PIECOST Approach", Defense Management Journal, Vol. 9, No. 3, July 1973, pp. 25–30.
232. Y. Gupta, "Life Cycle Cost Models and Associated Uncertainties", in Electronic Systems Effectiveness and Life Cycle Costing, Edited by J.K. Skwirzynski, Springer-Verlag, Berlin, 1983, pp. 535–549.
233. Y. Gupta, W.S. Chow, "Twenty-Five Years of Life Cycle Costing—Theory and Applications: A Survey", International Journal of Quality and Reliability Management, Vol. 2, No. 3, 1985, pp. 51–76.
234. E.C. Gustely, "Product Liability Loss Prevention: A Case History", ASQC Quality Congress Transactions, Boston, 1983, pp. 466–471.
235. F.M. Hall, "Economic Analysis for Data Base Management", Proceedings of the Annual Reliability and Maintainability Symposium, 1982, pp. 343–352.
236. J.F. Halldane, G. Meckler, "Constraints in Solar Life Cycle Cost Modelling", Proceedings of the Annual Meeting, American Section of

the International Solar Energy Society, V1, Orlando, Florida, June 6–19, 1977.
237. J.F. Halldane, G. Meckler and L.O. Degelman, "Parameter Grouping and Regionalizing to Simplify Solar Life Cycle Cost Models for Application to Solar Heating and Cooling of Buildings", Final Report. Meckler (Gershon) Associates, Inc., U.S., December 1976, 101 pp. NTIS No. TID-27681. Available from the National Technical Information Service (NTIS), Springfield, Virginia.
238. G.W. Hart, "DOD Share its Experience in Life Cycle Costing", Defense Management Journal, Vol. 14, No. 1, January 1978, pp. 16–18.
239. J.C. Harty, "A Practical Life Cycle Cost/Cost of Ownership Type Procurement Via Long Term/Multi Year, "Failure Free Warranty" (FFW) Showing Trial Procurement Results", Annals of Reliability and Maintainability, Volume 10, 1971, pp. 241–251.
240. G. Harvey, "Life Cycle Costing: A Review of the Technique", Management Accounting, October 1978, pp. 343–347.
241. D.P. Haworth, "The Principles of Life Cycle Costing", Industrialization Forum, Vol. 6, No. 3–4, 1975, pp. 13–20.
242. H. Hecht, "Economics of Reliability for Spacecraft Computers", Proceedings of Reliability and Maintainability Symposium, San Francisco, California, USA, January 1972, pp. 554–564.
243. D.A. Heiser, "New Techniques in R&M Contract Requirements", Proceedings of the Annual Reliability and Maintainability Symposium, 1986, pp. 480–485.
244. C.R. Heising, "Reliability and Maintenance Data Needed for High Voltage Circuit Breakers when Making Life Cycle Cost Studies", Proceedings of the 1979 Reliability Conference for the Electric Power Industry, 1979, pp. 103–108.
245. J.T. Henderson, "A Computerized LCC/ORLA Methodology", Proceedings of the Annual Reliability and Maintainability Symposium, 1979, pp. 51–55.
246. W. Hesse, "An Approach to Establishing the Most Economical Amount of Roof Insulation", ASHRAE Journal, March 1982, pp. 23–28.
247. C.V. Higbee, "Life Cycle Cost Analysis for Direct Use Geothermal Systems: An Introduction", Geothermal Energy, Vol. 12, No. 3, March 1984, pp. 7–8.
248. D. Hill and J. Wells, "Cost Effectiveness Guidelines for the Design of a System with Improved Reliability", Annals of Reliability and Maintainability, Vol. 4, 1965, pp. 53–62.
249. F.A. Hinrichs, "Army Views Challenge of Design to Cost", Defense Management Journal, Vol. 9, No. 3, July 1973, pp. 11–16.
250. C.J. Hitch, "Cost Considerations and System Effectiveness", Annals of Reliability and Maintainability, Vol. 4, 1965, pp. 1–7.

251. J.M. Hockett, et al., "Life Cycle Cost/Design-to-Cost Guidelines", Report No. M75-216, The MITRE Corp., Bedford, MA, 30 June 1975.
252. G.M. Hollander, "Life Cycle Cost Analysis—The Key to Economy", Consulting Engineer, Vol. 49, No. 3, September 1977, pp. 129–130.
253. G.M. Hollander, "Cost Studies and Engineering Economy: A Review of Terms", Specifying Engineer, September, 1976, pp. 78–81.
254. F.H. Hollister, R.R. Shorey, "Design to a Cost: Concept and Application", Defense Management Journal, Vol. 9, No. 3, July 1973, pp. 7–10.
255. Honeywell, Inc., "ANASN-101 Cost of Ownership Program", Document No. 371-1750, Honeywell Inc., Aerospace Division, St. Petersburg, Florida, March 1971.
256. J. Huetinck, M. Lipow and N.R. Garner, "Cost Effective Tradeoffs for Sequentially Deployed Systems", Proceedings of the Tenth Annual West Coast Reliability Symposium, February 21, 1969, pp. 1–20. Available from Western Periodicals Company, 13000 Raymer Street, North Hollywood, California 91605, U.S.A.
257. E. Huie and H.F. Harris, "Balanced Design-Minimum Cost Solution", in Design to Cost and Life Cycle Cost, AGARD Conference Proceedings No. 289, May 1980, pp. 18-1 to 18-8. Available from National Technical Information Service (NTIS), Springfield, Virginia, U.S.A.
258. K.K. Humphreys, P. Wellman, Basic Cost Engineering, Marcel Dekker, Inc., New York, U.S.A., 1987.
259. K.K. Humphreys (Editor), Project and Cost Engineers' Handbook, Marcel Dekker, Inc., New York, U.S.A., 1984.
260. R.B. Hutton, "Life Cycle Cost: Impact on Consumer Response to New Product Information for Major Home Appliances", Ph.D. Dissertation, University of Florida, 1977.
261. R. Hutton, W.L. Wilkie, "Life-Cycle Cost: Consumer Responses to New Information", Journal of Consumer Research, Vol. 6, March 1980, pp. 349–359.
262. R.B. Hutton, C.D. Anderson, "Life Cycle Costing: A Policy Alternative for Consumer Information", Changing Energy Use Futures, 2nd International Conference on Energy Use Management, Los Angeles, California, USA, October 22–26, 1979. Published by Pergamon Press, New York, NY, USA, and Oxford, England, 1979, vol. 2. pp. 741–750.
263. W.A. Hyman, D.J. Hughes, "Computer Model for Life-Cycle Cost Analysis of Statewide Bridge Repair and Replacement Needs", Transportation Research Record, No. 899, 1983, pp. 52–61.
264. J.L. Hysom, "Life Cycle Costing in Energy Conservation Analyses", Journal of Property Mangement, Vol. 44, November/December 1979, pp. 332–337.
265. Industrialization Forum, Vol. 6, 1979, (entire issue devoted to life cycle costing).

266. W. Ireland, "Cost Savings by Establishing Life Limits", Proceedings of the Annual Reliability and Maintainability Symposium, 1984, pp. 72–77.
267. G.C. Jackson and T.H. Maze, "Vehicle Life-Cycle Costing with Probabilistic Part Replacement and Repair Options", Transportation Research Record, No. 912, 1980, pp. 15–18.
268. A.K. Jeffrey, "Maximizing Logistics Return on R&D Investment", Proceedings of the Annual Reliability and Maintainability Symposium, 1987, pp. 258–263.
269. F.C. Jelen, C.L. Yaws, "How Energy Affects Life Cycle Costs", Hydrocarbon Processing, Vol. 56, No. 7, July 1977, pp. 89–91.
270. J.H. Johnson, "Life Cycle Costing of Hermetic Motors for Air Conditioning Applications", Conference Record, Paper presented at the 14th IAS Annual Meeting, Cleveland, Ohio, September 30–October 5, 1979. Published by IEEE (Catalogue No. 79CH1484-51A), New York, NY, 1979, pp. 542–549.
271. R.F. Johnson, "Cost Effective Spares Provisioning Models for Airline Operations", Proceedings of the Tenth Annual West Coast Reliability Symposium, Century City, California, February 21, 1969, pp. 165–216. Available from Western Periodicals Company, 13000 Raymer Street, North Hollywood, California 91605, U.S.A.
272. J.M. Jones, "Evolution of Techniques for LCC Analysis", in Design to Cost and Life Cycle Cost, AGARD Conference Proceedings No. 289, May 1980, pp. 7-1 to 7-13. Available from National Technical Information Service (NTIS), 5285 Port Royal Road, Springfield, Virginia 22161, U.S.A.
273. V.C. Jordan, "Life Cycle Costing—Introduction", Proceedings of the 10th Reliability and Maintainability Conference, 1971, p. 224.
274. M. Kamins, "Rules for Planned Replacement of Aircraft and Missile Parts", Report No. RM-2810-PR, The Rand Corporation, Santa Monica, California, November 1961.
275. R.J. Kapsch, "Life Cycle Costing Techniques Applicable to Law Enforcement Facilities", Law Enforcement Standards Program Report No: LESP-RPT-080/00, October 1974. Available from the Superintendent of Documents, U.S. Government Printing Office, Washington, D.C. 20402, U.S.A.
276. G. Kasouf, "LCC Versus Confidence Testing During Long Term Storage", Proceedings of the Annual Reliability and Maintainability Symposium, 1981, pp. 101–107.
277. I. Katz, R.E. Cavender, "Weapon System Life Cycle Costing", Annals of Reliability and Maintainability, Vol. 10, 1971, pp. 225–230.
278. R.J. Kaufman, "Life Cycle Costing for Capital Equipment Decisions", Automation, March 1969, pp. 75–80.
279. R.J. Kaufman, "Life Cycle Costing: Decision Making Tool Capital

Equipment Acquisitions", Journal of Purchasing, August 1969, Vol. 5, No. 3, pp. 16–31.
280. R.J. Kaufman, "Life Cycle Costing for Capital Equipment Decisions", Automotion, March 1969, pp. 75–80.
281. J.P. Keefe, "Life Cycle Costing in an Energy Crisis Era", Professional Engineer, July 1974, pp. 26–29.
282. P. Kelsey, "Florida Law Requires Life Cycle Costs on State-Financed Buildings", Air Conditioning, Heating and Refrigeration News, Vol. 5, 1975, pp. 80–85.
283. K.G. Associates, "Life Cycle Cost-Benefit Analysis: A Basic Course in Economic Decision Making", Report, Department of Health, Education and Welfare, Washington, DC, March 1976.
284. T.D. Kiang, "Life Cycle Costing—A New Dimension for Reliability Engineering Challenge", American Society for Quality Control (ASQC) Technical Conference Transactions, 1976, pp. 307–314.
285. T.D. Kiang, "Life Cycle Management Cost Model", Proceedings of the Annual Reliability and Maintainability Symposium, 1978, pp. 254–259.
286. I.C. Kidd, Jr., "Happiness is Quality and Cost Effectiveness", Defense Management Journal, Vol. 9, No. 3, July 1973, pp. 31–34.
287. P.S. Kilpatrick, A.L. Jones, "Life Cycle Cost Comparisons of Avionic System Design Alternatives", Proceedings of the IEEE 1974 National Aerospace and Electronics Conference (NAECON'74), Dayton convention Center, May 13–15, 1974, pp. 514–520.
288. W.G. Kindig, R.F. Bannecker, "Reliability Pay Off", Microelectronics and Reliability, Vol. 20, No. 1–2, January/February 1980, pp. 101–108.
289. R.J. Kinnier, "HVAC Energy Recovery Bin Method and Life Cycle Costing", Energy Engineering, Vol. 78, No. 2, 1981, pp. 17–33.
290. S.J. Kirk, "Life Cycle Costing: Problem Solver for Engineers", Specifying Engineering, Vol. 41, No. 6, 1979, pp. 123–129.
291. P.J. Klass, "DOD Stressing Life Cycle Costing Plan", Aviation Week and Space Technology, January 1967, pp. 33–34.
292. J. Klassen, "Life Cycle Cost Effectiveness", Heat Piping Air Conditioning, Vol. 58, No. 9, September 1986, pp. 75–84.
293. J. Kogan, "Life-Cycle Cost of Lifting Appliances", Proceedings of the Annual Reliability and Maintainability Symposium, 1979, pp. 28–33.
294. W.J. Kolarik, "Analysis Theory and Procedures for Determining and Predicting Availability Cost, and Intangible Effects for Farm Machinery Systems", Unpublished Ph.D. Dissertation, Oklahoma State University, USA, 1977.
295. W.T. Kolarik, "Life Cycle Costing and Associated Models", AIIE, Proceedings of the American Institute of Industrial Engineers (AIIE) Spring Annual Conference, 1980, pp. 59–64.
296. T.P. Konen, "Life Cycle Costing for Plumbing Systems", Specifications Engineering, Vol. 52, No. 1, July 1984, pp. 91–94.

297. T.P. Konen, "Life Cycle Cost Analysis for Plumbing Engineers", Plumbing Engineering, Vol. 10, No. 5, September–October 1982, p. 7.
298. D. Koral (Editor), "Three Government Buildings, Performance Specified, Energy-Integrated And Life-Cyle Costed, Point the Way To The Future Mode of Construction", Building Systems Design, Vol. 71, No. 6, October–November 1974, pp. 4–10.
299. A.W. Kratzke, J. Gauger, "Cost Optimization Subject to Availability Constraints", Proceedings of the Annual Reliability and Maintainability Symposium, 1979, pp. 73–79.
300. I.C. Kidd, Jr., "Happiness is Quality and Cost Effectiveness", Defense Management Journal, Vol. 9, No. 3, July 1973, pp. 31–34.
301. R.E. Kuehn, "R&M Applications of Cost-Benefit Analysis", Proceedings of the Annual Reliability And Maintainability Symposium, 1986, pp. 486–488.
302. S. Kyriakides, "Improving Quality and Efficiency of Military Tires for Low Life Cycle Costing", Proceedings of the International Rubber Conference, IRC 79, Italy, October 3–6, 1979. Available from International Rubber Conference Secretariat, Milan, Italy, 1979, pp. 684–693.
303. W.E. Lamar, "A Review and Assessment of System Cost Reduction Activities", AGARD (Advisory Group for Aerospace Research and Development) Conference Proceedings No. 289, Design to Cost and Life Cycle Cost, 1980, pp. 4-1 to 4-38. Available from National Technical Information Service (NTIS), Springfield, Virginia, U.S.A.
304. G. Lange et al., "Life Cycle Costing: Problems, Policies and Prospects", Army Procurement Research Office, Institute of Logistics Research, United States Army Logistics Management Center, Fort Lee, Virginia, USA, March 1970.
305. C.H. Langwost, "Life Cycle Costing Applied in Evaluating Alternative Short Take-off And Landing (STOL) Aircraft and Tracked Air Cushion Vehicles (TACV) as Modes of Transportation", Annals of Reliability and Maintainability, Vol. 10, June 27–30, 1971, pp. 231–235.
306. D.K. Lansdowne and Partners, "Life Cycle Costing: A Study", Report, David K. Lansdowne and Partners, 526 Newbold Street, London, Ontario N6E 1K6, Canada, 1976, 82 pp.
307. R. Laskin, and W.L. Smithhisler, "The Economics of Standard Electronic Packaging", Proceedings of the Annual Reliability and Maintainability Symposium, 1979, pp. 67–72.
308. L. Laulainen, "A Life Cycle Costing Model for Small Wind Energy Conversion Systems", Presented at the American Section of the International Solar Energy Society Wind Power: Energy Alternative for Midwest 2nd Conference, Minnesota, April 3–4, 1981, pp. 41–45.
309. P.E. Lauver, "Small Propulsion Engine LCC Considerations", Proceedings of the Annual Reliability and Maintainability Symposium, 1985, pp. 348–353.

310. M. Lenow, "Helicopter/Tilt Rotor Total System Cost Analysis Models 159 and 160", Report No. D8-2059, Vertol Division, The Boeing Aerospace Company, Seattle, March 1968.
311. M. Levine, D. Goldstein, I. Turierl, H. Herring and H. Estrada, "Life-Cycle Analysis in Residential Buildings and Consumer Appliances", Proceedings of the 15th Inter Soc. Energy Conversion Eng. Conference, Seattle, USA, Vol. 3, August 1980, pp. 2356–2362.
312. R.B. Lewis II, E.P. Laughlin, F.E. Spring, "U.S. Army Design-to-Cost Experience", in Design to Cost and Life Cycle Cost, AGARD Conference Proceedings No. 289, May 1980, pp. 3-1 to 3-11. Available from National Technical Information Service, Springfield, Virginia, U.S.A.
313. W.M. Lewis, Jr., "A Simple Statistical Method of Presenting the Uncertainty Associated with Life Cycle Cost Estimates", Study Report, Defense Systems Management College, Fort Belvoir, Virginia, USA, 1973.
314. B.S. Liebesman, "Reliability Requirements and Contractual Provisions", Proceedings of the Annual Reliability and Maintainability Symposium, 1985, pp. 1–5.
315. "Life Cycle Costing Emphasizing Energy Conservation: Guidelines for Investment Analysis", Reynolds, Smith and Hills, U.S., May 1977, 116 pp., NTIS No. ERDA-76/130 (Rev.), Available from the National Technical Information Service (NTIS), Springfield, Virginia, USA.
316. H.H. Lin, W. Kuo, "Reliability Cost in Software Life-Cycle Models", Proceedings of the Annual Reliability and Maintainability Symposium, 1987, pp. 364–368.
317. R.A. Lincoln, "How to Avoid Common Errors in Life-Cycle Cost Comparison", Power, Vol. 125, 1981, pp. 73–75.
318. M.O. Locks, "Maintainability and Life-Cycle Costing", Proceedings of the Annual Reliability and Maintainability Symposium, 1978, pp. 251–253. Available from IEEE, New York, U.S.A.
319. R.M. Lockerd, "Electron Technology Progress and Life Cycle Support", Defense Systems Management Review, Vol. 1, No. 2, 1977, pp. 1–11.
320. Logistics Management Institute, "Life Cycle Costing in Equipment Procurement", LMI Task 4c-5 Report, April 1965, Washington, D.C.
321. Logistics Management Institute, "Life Cycle Costs Procurement of Electric Ranges", Working Paper No. NBS-GCR-ETIP 76-79, 1976, Logistics Management Institute, Washington, D.C., U.S.A.
322. Logistics Management Institute, "Life Cycle Costing in the Procurement of Refrigerator-Freezers", Report, Logistics Management Institute, Washington, D.C., U.S.A., March 1976.
323. Logistics Management Institute, "Life Cycle Costing in Industry",

Task 67-21, Contract No. SD-271-74, Logistics Management Institute, Washington, D.C., USA, September 1967.

324. Logistics Management Institute, "Case Studies: Life Cycle Costing in Equipment Procurement", A Report to the Department of Defense, Washington, D.C., USA, July 1969.
325. Logistics Management Institute, "Life Cycle Costing in Equipment Procurement", A Report to the Department of Defense, Washington, D.C., April, 1965.
326. Logistics Management Institute, "Life Cycle Costing in Equipment Procurement—Supplemental Report", A Report to the Department of Defense, Washington, D.C., February 1967.
327. Logistics Management Institute, "Life-cycle Costing in System Acquisition", A Report to the Department of Defense, Washington, D.C., November 1969.
328. Logistics Management Institute, "Criteria for Evaluating Weapon System Reliability, Availability and Costs", Task 73-11, Logistics Management Institute, Washington, D.C., March 1974.
329. C.G. Love, R.A. Mathias, and G. Trebbe, "Minimization of Life-Cycle Costs of a Health Care System", Paper Prepared by Westinghouse Research and Development Center, Presented at the Systems Science and Cybernetics Conference (IEEE), Pittsburgh, Pennsylvania, USA, October 1970.
330. R.M. Loveless, "Using a Prediction Model to Reduce Test Program Set Costs", AUTOTEST CON Proceedings, Washington, D.C., November 1980, pp. 30–50.
331. R.T. Lund, "Life Cycle Costing as a Societal Instrument", Working Paper No. 77-15, Center for Policy Alternatives, Massachusetts Institute of Technology (MIT), Massachusetts, July 1977.
332. R.T. Lund, W.M. Denney, "Providing Individualized Consumer Information in the Market Place: A Life Cycle Cost Approach for Durable Products", Report, Center for Policy Alternatives, Massachusetts Institute of Technology, Massachusetts, December 1977.
333. L. Lynn, R.F. Howland, "Use Capital Ratio", in Modern Cost-Engineering Techniques, Edited by Herbert Popper, McGraw-Hill Book Company, New York, USA, 1970, pp. 367–372.
334. P.R. MacDiarmid, A.D. Pettinato, B.G. Johnson, "R/M/LCC Effects of Commercial Off-the-Shelf Equipment", Proceedings of the Annual Reliability and Maintainability Symposium, 1982, pp. 40–46.
335. P.E. Manzo, "Life-Cycle Costing", Specifying Engineer, September 1976, pp. 71–78.
336. O. Markowitz, "Case Histories of Life-Cycle Costing Procurement", Report, US Navy Aviation Supply Office, Philadelphia, Pennyslvania, USA, 1974.
337. K.E. Marks, H.G. Massey, B.D. Bradley, "An Appraisal of Models

Used in Life Cycle Cost Estimation for USAF Aircraft System", Report No. R-2287-AF, October 1978, The Rand Corporation, Santa Monica, California.

338. H.E. Marshall, "Simplified Energy Design Economics: Principles of Economics Applied to Energy Conservation and Solar Energy Investments in Buildings", National Bureau of Standards (NBS), Special Publication 544, 1980, 50 pp., Washington, D.C.
339. H.E. Marshall, S.R. Petersen, "Economics and the Selection and Development of Energy Standards for Buildings", Energy and Buildings, Vol. 2, 1979, pp. 89–99.
340. H.E. Marshall, R.T. Ruegg, "Energy Conservation in Buildings: An Economics Guidebook for Investment Decisions", National Bureau of Standard Handbook 132, May 1980, 149 pp., Washington, D.C.
341. H.E. Marshall, R.T. Ruegg, "Energy Conservation in Buildings Through Life-Cycle Costing", National Bureau of Standards, U.S., February 1977, 10 pp., Washington, D.C., NTIS No. PB-271 843/5ST. Available from the National Technical Information Service (NTIS), Springfield, Virginia.
342. H.E. Marshall, R.T. Reugg, "Life Cycle Costing Guide for Energy Conservation in Buildings", National Bureau of Standards, Washington D.C., 1979, 20 p. NTIS No. PB80-119084. Available from the National Technical Information Service, Springfield, Virginia.
343. J.K. Martin, D.L. O'Neal, "Energy and Cost Analysis of Residential Heat Pumps in Northern Climates", Proceedings of the 15th Inter Soc. Energy Conversion Engineering Conference, Seattle, USA, Vol. 3, August 1980, pp. 1802–1807.
344. J.E. Mayben, "Assurance of Availability and Life Cycle Costs", American Society for Quality Control (ASQC) Quality Congress Transactions, Detroit, 1982, pp. 465–474.
345. J.F. McCarthy, Jr., S.J. Novak, "Design-to-cost and Life Cycle Costing in Aerospace Industry", Industrialization Forum, Vol. 6, No. 3–4, 1975, pp. 37–51.
346. H.A. McClellan, R.D. Dighton, "The F/A-18 Hornet-Low Life Cycle Cost by Design", Proceedings of the Annual Reliability and Maintainability Symposium, 1983, pp. 403–410.
347. L. McClure, "Life Cycle Costing: A Selected Bibliography", Report No. RB 330-1, Martin Marietta Aerospace Corporation, Florida, USA, 1976.
348. N.B. McEachron, D.C. Hall, L.F. Lewis, "Life Cycle Costing as a Method of Procurement: A Framework and Example", Energy, Vol. 3, 1978, pp. 461–478.
349. N.B. McEachron, D.C. Hall, L.F. Lewis, "An Exploratory Evaluation: Assessing the Results of the LCC Water Heater Experiment, Vol. II: Appendices, P-Al, Prepared for the Federal Supply Service, General

Services Administration, Sponsored by the Experimental Technology Incentives Program, National Bureau of Standards, SRI International, Menlo Park, California, (September 1977).

350. N.B. McEachron, S.M. Royce, "The LCC Room Air Conditioner and Frostless Refrigerator/Freezer Experiments: A Joint Evaluability Assessment", Unpublished Draft Prepared for the Federal Supply Service and the Experimental Technology Incentives Program, SRI International, Menlo Park, California, September, 1977.
351. R.B. McLeod, "Requirements of an Economic Approach to Maintenance", IEEE Transactions on Engineering Management, Vol. EM-20, No. 3, August 1973, pp. 75–80.
352. G.R. McNichols, "Cost-Risk Procedures for Weapon System Risk Analysis", Proceedings of the Annual Reliability and Maintainability Symposium, 1981, pp. 86–94.
353. G.R. McNichols, "Treatment of Uncertainty in Life Cycle Costing", Proceedings of the Annual Reliability and Maintainability Symposium, 1971, pp. 56–61.
354. G.R. McNichols, "Cost-Risk Methodology for the V/STOL Acquisition Strategy Study", Report No. TR-7924-1, Management Consulting & Research, Inc., Falls Church, VA, USA, March 1980.
355. G.R. McNichols, "Uncertainties of LCC Predictions", NATO ASI Series, Series F: Computer and Systems Sciences, No. 3, Edited by J.K. Skwirzynski, Springer-Verlag, Munich, 1983, pp. 583–598.
356. R.J. McNichols, G.H. Messer, Jr., "A Cost-Based Availability Allocation Algorithm", IEEE Transactions on Reliability, Vol. R-20, No. 3, 1971, pp. 178–182.
357. G. Meckler, "Building Performande with Minimal Life Cycle Cost", Construction Specifier, Vol. 30, No. 5, May 1977, pp. 36–58.
358. G. Meckler, "An Integrated Life Cycle Costing and Energy Model", Industrialization Forum, Vol. 6, No. (3–4), 1975, pp. 9–12.
359. G. Meckler, M. Lockmanhekim, "Integrating Life-Cycle Cost into a Total Energy Analysis", Specifying Engineer, Vol. 37, No. 1, January 1977, pp. 71–76.
360. C.E. Meece, C.E. Spaeth, "Damage Tolerant Design—An Approach to Reducing the Life Cycle Cost of Gas Turbine Engine Disks", Proceedings of the American Institute of Astronautics and Aeronautics (AIAA), Society of Automotive Engineers (SAE) and American Society of Mechanical Engineers (ASME) Joint Propulsion Conference, Las Vegas, Nevada, USA, June 18–20, 1979, 6 pages.
361. T.D. Meitzler, R.M. Genet, "Life Cycle Cost Model for Inertial Systems", Report, Aerospace Guidance and Metrology Center, Newark, Ohio, June 1975.
362. W.C. Messecar, "Role of Life Cycle Costing in Fleet Planning Decisions", Annals of Reliability and Maintainability, Vol. 10, 1971, pp. 236–240.

363. H.S. Mickley, E.I. Korchak, "How to Find Optimum Economic Size for Multistage Heat Exchangers", in Modern Cost Engineering Techniques, McGraw-Hill Book Company, New York, 1970, pp. 464–467.
364. J.L. Miranda, "Determination if the Application of Life Cycle Costing Procedures to the Acquisition of US Government Vehicles would Result in Significant Saving", Term Research Report in Procurement and Acquisition Management, School of Systems and Logistics, Air Force Institute of Technology, Wright Patterson AFB, Ohio, USA, January 1972.
365. G.M. Montag, "Life Cycle Cost Analysis Versus Payback for Evaluating Project Alternatives", Heat Piping Air Conditioning, Vol. 56, No. 9, September 1984, pp. 75–78.
366. G.M. Montag, "Life Cycle Costing Building Replacement Analysis Model", ASHRAE Journal, Vol. 23, No. 8, August 1981, pp. 39–44.
367. G.M. Montag, "A Commercial Building Ownership Energy Conservation Cost Analysis Model", ASHRAE Journal Vol. 21, No. 6, 1979, pp. 49–52.
368. G.M. Montag, G.E. Lamp, "Energy-Conservation Evaluation—Rate of Return Versus the Present Worth Method of Life-Cycle Costing", ASHRAE Journal, Vol. 23, No. 11, 1981, pp. 52–55.
369. D. Monteith, B. Shaw, "Improved R, M, and LCC for Switching Power Supplies", Proceedings of the Annual Reliability and Maintainability Symposium, 1979, pp. 262–265. Available from IEEE, New York, U.S.A.
370. T.K. Moore, "Corrosion and Fatigue: Problems in Life Cycle Costing", Defense Management Journal, Vol. 12, No. 1, January 1976, pp. 48–53.
371. E.B. Morrison, "Will Life Cycle Costing Rule Out the Low Bidder", American Society for Quality Control (ASQC) Technical Conference Transactions, Los Angeles, California, 1969, pp. 343–347.
372. V.O. Muglia, A.S. Cici, R.H. Waln, "Optimum Life Cycle Costing", Proceedings of the Reliablity and Maintainability Symposium, 1974, pp. 369–376, Available from IEEE, New York, U.S.A.
373. H.A. Multhaup, "Design for Reliablity and Maintainability Life Cycle Cost Minimization", Proceedings of the 7th Annual Reliability Conference for the Electric Power Industry, Madison, Wisconsin, April 1980, pp. 88–92.
374. J.B. Munson, "Software Maintainability: A Practical Concern for Life-Cycle Costs", Computer, Vol. 14, No. 11, November 1981, pp. 103–109.
375. E.L. Murphree, Jr., "Economic Analysis Models for Evaluating Costs of a Life Cycle Cost Data Base", Technical Report, US Army Corp. Engineering Construction, Engineering Research Laboratory, September 1984, 57 pp.

376. B.A. Murphy, "Life-Cycle Cost Analysis: The Economics of Energy Conservation", ASHRAE Journal, Vol. 19, No. 10, October 1977, pp. 38–42.
377. K. Nagatani, "Life Cycle Saving: Theory and Fact", American Economic Review, Vol. 62, 1972, pp. 353–355.
378. I. Nathan, "A Reliability Cost-Benefit Analysis Study", Proceedings of the Annual Reliability and Maintainability Symposium, 1972, pp. 308–318.
379. I. Nathan, "Management Decision Utilizing Cost-Effectiveness Modeling", IEEE Transactions on Reliability, Vol. R-18, 1969, pp. 54–63.
380. National Aeronautics and Space Administration, "Methodology for Design-to-Cost of Aero Engines", in the Application of Design to Cost And Life Cycle Cost to Aircraft Engines", AGARD Lecture Series No. 107, May 1980, pp. B-1 to B-24. Available from National Technical Information Service (NTIS), Springfield, Virginia, U.S.A.
381. J.R. Nelson, "Life-Cycle Analysis of Aircraft Turbine Engines", Report No. R-2103-AF, November, 1977, The Rand Corporation, Santa Monica, California.
382. J.R. Nelson, P. Knoske-Dey, M.R. Fiorello, J.R. Gebman, G.K. Smith and A. Sweetland, "A Weapon-System Life-Cycle Overview: The A-7D Experience", Report No. R-1452-PR, October 1974, The Rand Corporation, Santa Monica, California.
383. J.R. Nelson, "An Approach to the Life-Cycle Analysis of Aircraft Turbine Engines", in the Application of Design to Cost and Life Cycle Cost to Aircraft Engines, AGARD Lecture Series No. 107, May 1980, pp. 2-1 to 2-27. Available from National Technical Information Service (NTIS), Springfield, Virginia, U.S.A.
384. B. De Neumann, "Life Cycle Cost Models", Proceedings of the NATO Advanced Study Institute on Electronic Systems Effectiveness and Life Cycle Costing, Norwich, United Kingdom, Edited by J.K. Skwirzynski, July 19–31, 1982, pp. 513–532.
385. W. Ng, "Life Cycle Costing with a Programmable Calculator", Heating, Piping, Air Conditioning, Vol. 53, No. 4, 1978, pp. 90–93.
386. L.A. Nieves, "Energy Performance Standard: The Economic Issues", ASHRAE Journal, Vol. 23, No. 1, January 1980, pp. 49–52.
387. J.W. Noah, "Concepts and Techniques for Summarizing Defense System Costs", Report, Franklin Institute, Washington, DC, 1965.
388. J. Noah and H.L. Eskew, "Cost-of-Ownership-Pitfalls", Proceedings of the Annual Reliability and Maintainability Symposium, 1975, pp. 221–224.
389. R.E. O'Donohue, Jr., "Life Cycle Costing", Commanders Digest, Vol. 1, 1975, pp. 2–5.
390. D.L. O'Neal, "Energy and Cost Analysis Residential Heating Systems," Report No. ORNL-ON-25, Oak Ridge National Laboratory, Tennessee, USA, July 1978.

391. P.G. Osterbeck, R.C. Sutton, L.D. Hawkins, "Introducing Cost Effectiveness into the Tactical Airplane Design in a Cost Effective Manner", 17th Aerospace Sciences Meeting, AIAA Paper No. 79-0235, January 1979. Available from the American Institute of Astronautics and Aeronautics (AIAA), New York.
392. V. Ostrander and D.E. Parker, "A Life Cycle Planning and Budgeting Model", Industrialization Forum, Vol. 6, No. 3–4, 1975, pp. 83–87.
393. R.F. Panella, M.A. Barga, R.G. McNally, "The Role of Advanced Technology on Turbine Engine Life Cycle Cost", in Design to Cost and Life Cycle Cost, AGARD Conference Proceedings No. 289, May 1980, pp. 26-1 to 26-13. Available from National Technical Information Service (NTIS), Springfield, Virginia, U.S.A.
394. R.F. Panella, "Engine Life Cycle Cost: A Laboratory View", Journal of Aircraft, February 1976, pp. 156–158.
395. E.T. Parascos, J.A. Arceri, "Reliability Engineering and Underground Equipment Failure, Cost and Manufacturer's Analysis", Proceedings of the Third Annual Reliability Engineering Conference for the Electric Power Industry, September 23–24, 1976, pp. 25–30.
396. C.S. Park, "Counting the Costs: New Measures of Manufacturing Performance", Mechanical Engineering, Vol. 109, No. 1, January 1987, pp. 66–71.
397. J.D. Patton, Jr., "Dollar-Based Specification of RAM", Proceedings of the Annual Reliability and Maintainability Symposium, 1979, pp. 18–22.
398. R.M. Paulson, R.B. Waina, L.H. Zacks, "Using Logistics Models in System Design and Early Support Planning", Report No. R-550-PR, The Rand Corporation, Santa Monica, California, February 1971.
399. G.F. Pearson, "Life-Cycle Costing in an Energy Crisis Era", Professional Engineer, Vol. 44, No. 7, July 1974, pp. 26–29.
400. R.J. Pearson, "Aerospace Systems Analysis Approach to Energy Conservation in Heating, Ventilating And Air Conditioning Systems", ASME Paper No. 79-EN AS-1, July 16–19, 1979, 8 pp. Available from the American Society of Mechanical Engineers, New York.
401. P.C. Pedrick, "Survey of Life Cycle Costing Practices of Non-Defense Industry", Proceedings of the 1968 Anual Symposium on Reliability, Boston, Massachusetts, January 16–18, 1968, pp. 188–192.
402. C.A. Pell, "A Reliability Incentive Based Bid Decision", Proceedings of the Annual Reliability and Maintainability Symposium, 1977, pp. 396–400.
403. S.S. Penner, M.R. Brambley, R. Schmalensee, "Life Cycle Costing for Consumers of Energy-Conserving Devices", Energy, Vol. 3, 1978, pp. 415–419.
404. G.S. Peratino, "Air Force Approach to Life Cycle Costing", Proceedings of the 1968 Annual Symposium on Reliability, Boston, Massachusetts, January 16–18, 1968, pp. 184–187.

405. B. Perkins, "Life Cycle Costing: An Approach to Method", Architectural Record, August 1975, pp. 53–59.
406. W.R. Perrigo, J.L. Easterday, "Avionic Equipment Reliability and Low Life Cycle Cost", Proceedings of the IEEE 1974 National Aerospace and Electronics Conference (NAECON'74), held at Convention Center, Dayton, Ohio, USA, May 1974, pp. 521–531.
407. E. Peterson, "Life Cycle System Cost Effectiveness", Proceedings of Annual Reliability and Maintainability Symposium, 1974, pp. 48–51.
408. E. Peterson, "Maintainability Application to System Effectiveness Quantification", Maintainability Bulletin, No. 15, Electronics Industries Association, USA, September 1969.
409. E. Peterson, "Irrationality of Marine Gas Turbine Ship Cost-Effective Analysis", Naval Engineers Journal, October 1970, pp. 25–34.
410. D.E. Peterson, "Life Cycle Cost Analysis of Pavements", Transportation Research Board, National Research Council, Washington, D.C., USA, 1985, 136 pp.
411. S.R. Petersen, Economic Analysis of Insulation in Selected Masonry and Wood-Frame Walls", National Bureau of Standards, U.S., Report No. NBSIR 79-1789, Washington, D.C., September 1979, 138 pp.
412. S.R. Petersen, "Economic Optimization in the Energy Conservation Design of Single-Family Housing", National Bureau of Standards, Washington, D.C., U.S., 1976, 13 pp., NTIS Report No. PB-280 584/4 ST. Available from the National Technical Information Service (NTIS), Springfield, Virginia.
413. S.R. Petersen, "Retrofitting Existing Housing for Energy Conservation: an Economic Analysis", National Bureau of Standards, Building Science Series 64, Washington D.C., December 1974, 76 pp.
414. S.R. Petersen, G.E. Kelly, D.A. Didion, "Economic Analysis of Improved Efficiency for Central Air Conditioners", June 1980, 55 pp. NTIS Report No. PB80-209885. Available from the National Technical Information Service (NTIS), Springfield, Virginia.
415. T. Pike, M. Golubov, "GHT Design Options—Cost Benefit Analysis", Proceedings of the 4th National Passive Solar Conference, Kansas City, Missouri, October 1979, pp. 407–410.
416. D.A. Pilati, "Room Air Conditioner Life Time Cost Considerations: Annual Operating Hours and Efficiencies", Report No. ORNL-NSF-EP-85, Oak Ridge National Laboratory, Tennessee, USA, October, 1975.
417. H.L. Platzer, "Sparing for Repairable, Redundant Systems", Proceedings of the Annual Reliability and Maintainability Symposium, 1983, pp. 394–397.
418. J.M. Porterfield, "Life Cycle Costing and Economic Ranking of Weatherization Options", Proceedings of 3rd National Conference, Technology on Energy Conservation, Tucson, Arizona, January 1979, pp. 508–517.

419. J.W. Powell, "Economic Evaluation of Passive Solar Designs for Urban Environments", Proceedings of the 4th National Passive Solar Conference, Kansas City, MO, October 3–5, 1979, pp. 1–5.
420. J.W. Powell, "An Economic Model for Passive Solar Designs in Commercial Environments", National Bureau of Standards, Building Science Series 125, Washington, D.C., June 1980, 150 pp.
421. J.W. Powell, R.T. Ruegg, "Life Cycle Costing of Solar Energy Investments", National Bureau of Standards, Washington, D.C., January 1981, 8 pp., NTIS Report No. PB81-197477. Available from the National Technical Information Service (NTIS), Springfield, Virginia.
422. J.T. Prohaska, W.G. Briggs, J.B. DeWolf, R.T. Lund, "Life-Cycle Costs: Concepts, Considerations, and Modelling", Section of Final Report on National Science Foundation (NSF) Grant to Center for Policy Alternatives, Massachusetts Institute of Technology (MIT), Massachusetts, USA, 1977.
423. L.H. Putnam, "A General Empirical Solution to the Macro Software Sizing and Estimating Problem", in Tutorial-Software Cost Estimating and Life-Cycle Control: Getting the Software Numbers, Computer Society Press, IEEE Computer Society, 1980, pp. 141–157.
424. L.H. Putnam, (Editor), "Software Cost Estimating and Life-Cycle Control", Computer Society Press, IEEE Computer Society, 1980, 349 pp.
425. A.S. Rakhra, "Building and Life-Cycle Costing", Canadian Building Digests CBD 212, National Research Council, Ottawa, Canada, July 1980, 4 pp.
426. M. Ralph, "Real Property Facility Costs as Related to the Life Cycle Process with Application to Building Structures", A Report to the Public Building Service of the General Services Administration, Washington, DC, February 1974.
427. M. Ralph, "A Life Cycle Concept for Buildings and Facilities", A Report, the Building Research Division, Institute of Applied Technology, National Bureau of Standards, Washington, DC, April 1970.
428. M. Ralph, "A Uniform Language and System for Cost Decisions, in the Building Process", Report, the Building Research Division, Institute for Applied Technology, National Bureau of Standards, Washington, DC, July 1971.
429. B. Rapp, J. Seimer, "A Model for the Evaluation of Energy Conservation Investments", Energy & Buildings, Vol. 2, 1979, pp. 203–208.
430. B.R. Rashkow, "Life-Cycle Costing and The Engineer", Specifying Engineering, vol. 56, No. 1, July 1986, pp. 87–91.
431. G.A. Raymond, "Reliability Versus the Cost of Failure", Proceedings of the Fourth National Symposium on Reliability and Quality Control, 1957, pp. 187–188.

432. RCA Communications Systems Division, "Total Life Cycle Cost of Ownership Study Report—Radio Set AN/URC-78 (XE-2)/V", Preliminary Report (Contract DAAB07-71-C-0182) to U.S. Army Electronics Command, Fort Monmouth, New Jersey, April 1972.
433. W.J. Rechmeyer, H.G. Shook, R.I. Houston, "Application of Integrated Logistics Support to Life Cycle Management of Material Development", Final Report, International Research Institute, Under Contract DAHC 19-71-C0032, Department of Army, USA, June 1972.
434. T.L. Regulinski, Y.P. Gupta, "Reliability Cost Estimation: Managerial Perspectives", IEEE Transactions on Reliability, Vol. R-32, No. 3, August 1983, pp. 276–281.
435. T.L. Regulinski, "Uncertainties in Monte Carlo Simulation of Life Cycle Costs", in Electronic Systems Effectiveness and Life Cycle Costing, Edited by J.K. Skwirzynski, Springer-Verlag, Berlin, 1983, pp. 599–620.
436. V. Rehg, "A Computer Simulation of the Life Cycle", Proceedings of the Annual Reliability and Maintainability Symposium, 1973, pp. 564–576.
437. V. Rehg, R. Sowar, "Visual Decision Making", Proceedings of the Reliability and Maintainability Symposium, January 1974, pp. 238–243.
438. H. Reiche, "Life Cycle Cost", in Reliability and Maintainability of Electronic Systems, edited by J.E. Arsenault and J.A. Roberts, Computer Science Press, Potomac, Maryland 20854, USA, 1980, pp. 3–23.
439. W.L. Reid, R.E. Shannon, "Cost-Analysis and Optimization Study for Solar Heating and Cooling Systems", Alabama University in Huntsville, U.S., January 1979, 51 pp. NTIS Report No. N79-20499/6ST. Available from the National Technical Information Service (NTIS), Springfield, Virginia.
440. A.J. Reiger, "Market Place Realities and Solar Economics", Sun: Mankinds Future Source of Energy: Proceedings of the International Solar Energy Society Congress: New Delhi, India, January 16–21, 1978, vol. 1, pp. 248–252.
441. D.B. Reister, "The Total Energy Cost of Freight Transport", Energy, Vol. 3, 1978, pp. 493–498.
442. D.S. Remer, S. Adnul-Ganity, K. Khan, "A Model for Life Cycle Cost Analysis with a Learning Curve", Engineering Economist, Vol. 27, No. 1, Fall 1981, pp. 29–58.
443. "Report on Building Life Costs", A Report Prepared by the Council of Ontario Universities, Toronto, Canada, November 1973.
444. S. Reynolds, F. Hills, "Life Cycle Cost Emphasizing Energy Conservation: Guidelines for Investment Analysis", Report, Energy Research and Development Administration (ERDA), Washington, DC, 1977.
445. S. Reynolds, F. Hills, "Life Cycle Costing Emphasizing Energy Conservation—Guidelines for Investment Analysis", Report, US Department

of Commerce, September 1976, Available from the National Technical Information Service, Springfield, Virginia.

446. C. Rich, "Using Life-Cycle Analysis to Improve Plant Design", Plant Engineering, Vol. 32, 1978, pp. 151–154.
447. P.W. Richardson, F.W. Miller, M.N. White, "Design, Performance, and Life Cycle Cost Relationships for a 500 kw Space Solar Array", Proc. 15th Intersoc. Energy Conversion Engineering Conference, Vol. 2, Seattle, Washington, USA, 1980, pp. 1396–1400.
448. H.S. Riddel, "Life Cycle Costing in the Chemical Industry: Two Case Studies", Terotechnica, Vol. 2, No. 1, January 1981, pp. 9–21.
449. J.L. Rise, "Establishing Life Cycle Cost Practices for an Electric Power Company", Proceedings of the 1983 International Reliability, Availability and Maintainability Conference for Electric Power Industry, pp. 221–225.
450. F. Roach, S. Noll, S. Ben-David, "Passive and Active Residential Solar Heating: Comparative Economic Analysis of Select Designs", Energy, Vol. 4, No. 4, August 1979, pp. 623–644.
451. J.L. Roca, "Approach in the Life Cycle Cost of Integrated Logistic Support", Microelectronics and Reliability, Vol. 27, No. 1, 1987, pp. 25–27.
452. J. Rose, E. Phelps, "Cost of Ownership Application to Airplane Design", Proceedings of the Annual Reliability and Maintainability Symposium, 1979, pp. 47–50.
453. R.G. Rose, "Design to Cost Viewed Against the Achievement of Optimum System Capability", in Design to Cost and Life Cycle Cost, AGARD Conference Proceedings No. 289, May 1980, pp. 24-1 to 24-6. Available from National Technical Information Service (NTIS), Springfield, Virginia, U.S.A.
454. H. Rosenberg, J.H. Witt, "Effects on LCC of Test Equipment Standardization", Proceedings of the Annual Reliability and Maintainability Symposium, 1976, pp. 287–292. Available from IEEE, New York, U.S.A.
455. R.A. Rosien, "Series 320 Radar Family—A Cost Effective Approach to Intertialess Neam Steering", EASCON Record/IEEE Electronic and Aerospace Systems Conversion Conference, Arlington, Virginia, September/October, 1980, pp. 182–189.
456. D. Rosoff, "The Background Progress and State-of-the-Art in Applying Life-Cycle Concepts in Buildings and Systems", Proceedings of the Conference on Improving Efficiency and Performance of HVAC Equipment and Systems for Commercial and Industrial Buildings, Ray W. Herrick Laboratories, Volume 1, Purdue University, Indiana, April 1976.
457. A.O. Rotz, "Warranty Pricing with a Life Cycle Cost Model", Proceedings of the Annual Reliability and Maintainability Symposium, 1986, pp. 454–456.

458. W.M. Rowell, "Design-to-Cost and the F-16 Multirole Fighter", in Design to Cost and Life Cycle Cost, AGARD Conference Proceedings No. 289, May 1980, pp. 9-1 to 9-14. Available from National Technical Information Service (NTIS), 5285 Port Royal Road, Springfield, Virginia 22161, U.S.A.
459. R.T. Ruegg, R.E. Chapman, "Economic Evaluation of Windows in Buildings", NBS Building Science Series 119, Report, US Department of Commerce, National Bureau of Standards, Washington, D.C., USA, April 1978.
460. R.T. Ruegg, "Evaluating Incentives for Solar Heating", Report, US Department of Commerce, National Bureau of Standards, Washington, D.C., September 1976.
461. R.T. Ruegg, "Life Cycle Costs and Solar Energy", ASHRAE Journal, Vol. 18, No. 11, November 1976, pp. 22–25.
462. R.T. Ruegg, "Solar Heating and Cooling in Buildings: Methods of Economic Evaluation", US Department of Commerce, National Bureau of Standards, Washington, D.C., July 1975, 47 pp. NTIS Report No. COM 75-11070. Available from the National Technical Information Service (NTIS), Springfield, Virginia.
463. R.T. Ruegg, J.S. McConnaughey, G. Thomas Sav, K.A. Hockenbery, Life-Cycle Costing: A Guide for Selecting Energy Conservation Projects for Public Buildings", NBS Building Science Series 113, September 1978. Available from Superintendent of Documents, Government Printing Office, Washington, D.C. 20402, U.S.A.
464. R.T. Ruegg, "Calculating the Solar Dollar Gains: ins and Outs of Life Cycle Costing", National Bureau of Standards, Washington, D.C., July 1979, 4 pp. NTIS Report No. PB80-132897, Available from the National Technical Information Service (NTIS), Springfield, Virginia.
465. R.T. Reugg, "Economic Feasibility of Solar Applications to Office Buildings and Retail Stores", Proceedings of Solar Energy Market Analysis and Evaluation Contractor's Review Meeting, Washington, DC, April 8–10, 1980, pp. 1–7.
466. R.T. Ruegg, "Life-cycle Cost Manual for the Federal Energy Management Program", National Bureau of Standards, Washington, D.C., December 1980, 240 pp. NTIS Report No. PB81-136269. Available from the National Technical Information Service, Springfield, Virginia.
467. R.T. Ruegg, "Life Cycle Costing Translating Cost-Saving Potential into Real Dollars in the Area of Building Energy Management", Building Operating Management, Vol. 26, No. 3, March 1979, pp. 52–56.
468. R.T. Ruegg, "Recommended Practice for Measuring Life Cycle Costs of Buildings and Building Systems", National Bureau of Standards, Washington, D.C., Report No. NBSI 80-240, June 1980, 77 pp. NTIS Report No. PB80-203649. Available from the National Technical Information Service (NTIS), Springfield, Virginia.

469. R.T. Ruegg, "Using Economic Analysis to Achieve Cost", Transaction of (AACE) American Association of Cost Engineers Annual Meeting, Milwaukee, Wisconsin, June 26–29, 1977, pp. 196–202.
470. S.H. Russell, M.K. Wees, "Life Cycle Costing for Resource Recovery Facilities", Proceedings of 9th National Waste Process Conference, Washington, DC, May 1980, pp. 110–115.
471. W.J. Ryan, "Procurement Views of Life Cycle Costing", Proceedings of the 1968 Annual Symposium on Reliability, Boston, Massachusetts, January 16–18, 1968, pp. 164–168.
472. G.L. Sadler, A. Constantinides, L. Philopson, "An Evaluation of Weapon System Life Cycle Analysis Models", Proceedings of the Annual Reliability and Maintainability Symposium, 1974, pp. 460–468.
473. G.L. Sadler, B.I. Rachowitz, "Design-to-Cost Methodology", 35th Annual Conference of Society of Allied Weight Engineers, Philadelphia, Pa., May 24–26, 1976, 23 pages. SAWE Paper 1138, 76/05/00, 77A12199.
474. L.S. Schlosser, "Practical Innovation can Mean Lower Life-Cycle Costs", Defense Management Journal, Vol. 14, No. 5, 1978, pp. 32–35.
475. H.E. Schock, Jr., "Regulatory Cost Management and Product Assurance", Proceedings of the Annual Reliability and Maintainability Symposium, 1979, pp. 13–17.
476. B.A. Schmidt, "Preparation for LCC Proposals & Contracts", Proceedings of the Annual Reliability and Maintainability Symposium, 1979, pp. 62–66.
477. W.N. Schmitz, "AV-8B-M by Design: Impact on Supportability", Proceedings of the Annual Reliability and Maintainability Symposium, 1987, pp. 240–246.
478. N. Schroeder, C. Sontz, "Life Cycle Cost Design Trade-Offs for a Developmental UHF Modular Transceiver", Proceedings of the IEEE 1974 National Aerospace and Electronics Conference (NAECON'74), Held at the Dayton Convention Center, Dayton, May 13–15, 1974, p. 533.
479. J.K. Seger, "Reliability Investment and Life-Cycle Cost", IEEE Transactions on Reliability, Vol. R-32, No. 3, August 1983, pp. 259–263.
480. J.K. Seger, "Repair-Discard Concepts in Design", Proceedings of the Annual Reliability and Maintainability Symposium, 1982, pp. 21–24.
481. K. Seiler, Introduction to System Cost-Effectiveness, John Wiley & Sons, New York, 1969.
482. M.R. Seldon, "Life Cycle Costing: A Better Method of Government Procurement", Westview Press, Boulder, Colorado, 1979.
483. L. Sessen, "Life Cycle Cost Testing", Evaluation Engineering, January/February 1974, pp. 80–85.
484. Y.S. Sherif, "Life Cycle Costing Methodology for the Assessment of Process Heat Generation by Solar Energy", Microelectronics and Reliability, Vol. 23, No. 6, 1983, pp. 1069–1074.

485. Y.S. Sherif, "Optimal Maintenance Model for Life Cycle Costing Analysis", Reliability Engineering, Vol. 3, No. 3, May 1982, pp. 173–177.
486. Y.S. Sherif and W.J. Kolarik, "Life Cycle Costing: Concept and Practice", OMEGA, Vol. 19, No. 3, 1981, pp. 287–296.
487. J.M. Sherman, M.S. Gresham, D.L. Fergason, Wind Systems Life Cycle Cost Analysis: A Description and Users Manual, Rockville Center, US Department of Energy, Windbooks, New York, 1983.
488. W.W. Shoemaker, "Life Cycle Cost Before, During and After the Design of Aircraft Propulsion Systems", ASME American Society of Mechanical Engineers Paper No. 81-GT-111, New York, March 1981, 11 pp.
489. W.W. Shoemaker, "Life Cycle Cost as a Tool in the Detail Design Advanced Propulsion Systems", 16th AIAA/SAE/ASME Jet Propulsion Conference, Hartford, Connecticut, USA, June–July, 1980, 9 pp.
490 R.R. Shorey, "Managing Downstream Weapons Acquisition Costs", Defense Management Journal, Vol. 12, No. 1, January 1976, pp. 10–18.
491. M.J. Shumaker, J.C. DuBuisson, "Tradeoff of Thermal Cycling VS Life Cycle Costs", Proceedings of the Annual Reliability and Maintainability Symposium, 1976, pp. 300–305.
492. D.P. Siewiorek, R.S. Swarz, "The Theory and Practice of Reliable System Design", Digital Equipment Corporation, 1982, pp. 11–16, 303–321.
493. R.L. Singleton, "Program Management Guidelines for Acquiring More Reliable and Lower Life-Cycle Cost Avionics Equipment", Report No. TR-1251-2-1, The Analytic Science Corporation, Reading, Massachusetts, 1978.
494. W.B. Sininger, "Engine Life Cycle Cost", American Institute of Aeronautics and Astronautics and Society of Automotive Engineers 12th Propulsion Conference, Palo Alto, California, U.S.A., July 26–29, 1976, 11 pages. AIAA Paper 76-754, 76/07/00, 76A38249.
495. B.D. Sivazlian, "Life Cycle Cost Analysis Under Uncertainty in Systems Life Time", The Engineering Economist, Vol. 25, No. 2, 1980, pp. 91–105.
496. J.K. Skwirzynski, (Editor), Electronic Systems Effectiveness and Life Cycle Costing, Springer-Verlag, New York, 1983, 732 pp.
497. R.E. Slesinger, "Cost Analysis Approach to Utility Making", Public Utilities Fortnightly, Vol. 86, No. 7, 1970, p. 56.
498. J.R. Smeed, "Life Cycle Costing in Air Conditioning Design", Australian Refrigeration, Air Conditioning and Heating, January 1980, pp. 10–19.
499. C.C. Smith, R.E. Dynes and K.A. Chorney, "Cost Effectiveness Methodology—Part I: Guidelines for System Life Cycle Costing", Technical Report No. RADC-TR-119, TRW Systems Group, Redondo Beach, California, July 1978.

500. I.M. Soi, K.K. Aggarwal, "A Life Cycle Cost Viewpoint of Software Maintainability", Computers and Electrical Engineering, Vol. 8, No. 4, December 1981, pp. 277–282.
501. J.P. Solomond, G.A. Marseglia, "Cost Optimizing System to Evaluate Reliability (COSTER)", Proceedings of the Annual Reliability and Maintainability Symposium, 1977, pp. 385–390.
502. G.G. Somers, W.D. Wood (Editors), Cost-Benefit Analysis of Man Power Policies, Report Industrial Relations Centre, Queen's University, Kingston, Ontario, Canada, 1969, 272 pp.
503. H. Southworth, Jr., "A Cost Relevance Tree Procedure for Allocating Development Expenditure", IEEE Transactions on Engineering Management, Vol. EM-20, No. 1, February, 1973.
504. F.J. Spayth, J.F. Keathing, J.G. Taylor, "Ultra Reliable VHF-FM Receiver Transmitter Life Cycle Cost of Ownership Study", Research and Development Technical Report ECOM-0127-10, US Army Electronics Command, Fort Monmouth, New Jersey 1971.
505. R.M. Standahar (Editor), Proceedings of OSD Aircraft Engine Design and Life Cycle Cost Seminar, Naval Air Development Center, U.S.A., May 1978.
506. F.W. Stanley, "Factors to Consider when Selecting Solar-Energy Systems: Life Cycle Costing Basics", Plant Engineering, Vol. 33, No. 19, 1979, pp. 159–161.
507. J.W. Stansberry, "Source Selection and Contracting Approach to Life Cycle Cost Management", Defense Management Journal, Vol. 12, No. 1, January 1976, pp. 19–22.
508. R.B. Stauffer, "Collection of Data for Estimating the Probable Life Cycle Costs of Solar Energy Systems", Proceedings of the DOE Performance Monitoring Techniques for Evaluation of Solar Heating and Cooling Systems Conference, Washington, D.C., April 3–4, 1978, pp. 29–34.
509. R.E. Steere, E.G. Koepnick, R.A. Dean, "Progress on the US Air Force Approach for the Practical Management of Engine Life Cycle Costs," AGARD Lecture Series No. 107, The Application of Design to Cost and Life Cycle Cost to Aircraft Engines, May 1980, pp. 4-1 to 4-10. Available from National Technical Information Service (NTIS), Springfield, Virginia, U.S.A.
510. F. Stehle, P.R. Oyerly, A. Scarfile, "The Development and Application of a Cost-Effectiveness Assessment Procedure for Shipboard Electronic Equipments", Annals of Reliability and Maintainability, Vol. 5, 1966, pp. 61–74.
511. J.D. Stelle, "Use of Models in Support of Policies to Control Avionics Life Cycle Cost", Report No. WN-8236-ARPA, The Rand Corporation, Santa Monica, California, May 1973.

512. P.E. Stephan, "Human Capital Production: Life Cycle Production with Different Learning Technologies", Economic Inquiry, Vol. 14, No. 4, 1976, pp. 539–557.
513. M.O. Stern, "Life Cycle Costing, Government Policies and the Diffusion of Energy-Conserving Technology", Energy, Vol. 3, No. 2, April 1978, pp. 173–202.
514. D. Sternlight, "Life Cycle Costing as an Economic Design and Decision Tool", Naval Research Logistics Quarterly, Vol. 18, No. 4, 1971.
515. S.W. Stiefel, P.C. Goodman, W.B. Beine, "Application of Life-Cycle Costing to Hand-Held Dryer: A Field Demonstration for Small Appliances", US Department of Commerce, National Bureau of Standards, Washington, D.C., 1978.
516. S.W. Stiefel, P.C. Goodman, T.J. Wang, A.C. Lewis, "Life Cycle Costing for Major Household Appliances: An Analysis of Ownership Cost Variations", Report, The US Department of Energy, Washington, DC, July 1979.
517. S.W. Stiefel, K.S. Justin, M. Howard, "Life Cycle Costing: An Assessment of Practicability for Consumer Products", Report, Center for Consumer Products Technology, National Bureau of Standards, Washington, DC, 1976.
518. R.G. Stokes, F.N. Stehle, "Some Life Cycle Cost Estimates for Electronic Equipments: Methods and Results", Proceedings of the 1968 Annual Symposium on Reliability, Boston, Massachusetts, January 16–18, 1968, pp. 169–183.
519. N.C. Stordahl, J.L. Short, "The Impact and Structure of Life Cycle Costing", Proceedings of the Annual Symposium on Reliability, Boston, Massachusetts, January 16–18, 1968, pp. 509–515.
520. Stran-Steel Corporation, "New Life Cycle Construction Costing: How to Plan Lifetime Savings into Your Building Investment", Houston, Texas, USA, 1974.
521. A. Stratton, "The Principles and Objectives of Cost-Effectiveness Analysis", The Aeronautical Journal of the Royal Aeronautical Society, Vol. 72, 1968, pp. 43–53.
522. D. Stuhlbarg, "Economic Justification for Equipment", in Modern Cost-Engineering Techniques, Edited by Herbert Popper and the Staff of Chemical Engineering, McGraw-Hill Book Company, New York, USA, 1970, pp. 362–366.
523. K. Surendran, K.R. Chandra, "Cost Reliability Trade-off in an Electronic Module, "Microelectronics and Reliability, Vol. 15, 1976, pp. 493–496.
524. W.C. Swan, D.W. Bouwer, F.F. Tolle, "Life Cycle Cost Impact on Design Considerations for Civil Transport Aircraft Propulsion

Systems", Proceedings of the Third International Symposium on Air Breathing Engines, Munich, West Germany, March 7–12, 1976, pp. 11–26.

525. R.J. Symon, K.J. Dangerfield, "The Application of Design to Cost at Rolls-Royce", in the Application of Design to Cost and Life Cycle Cost to Aircraft Engines, AGARD Lecture Series No. 107, May 1980, pp. 7-1 to 7-17. Available from National Technical Information Service (NTIS), Springfield, Virginia, U.S.A.
526. R. Tassinari, "Organizing a Design-to-Cost Program", in Design to Cost and Life Cycle Cost, AGARD Conference Proceedings No. 289, May 1980, pp. 13-1 to 13-14. Available from National Technical Information Service, Springfield, Virginia, U.S.A.
527. J.H. Taylor, "A Realistic Approach to System Life Cycle Cost", Proceedings of IEEE 1974 National Aerospace and Electronics Conference (NAECON 74), 1974, pp. 539–546.
528. J.H. Taylor, "AN/ASM-101 Gimbaled Electrostatic Gyro Aircraft Navigation System Logistics Support Costs and Life Cycle for Various Quantities and Support Alternatives", Document No. FS-LOG-101, Honeywell Inc., Aerospace Division, St. Petersburg, Florida, July 1973.
529. J.H. Taylor, "Industrial Prospective on Inertial Navigation System Life Cycle Cost", Conference Proceedings Book II—Life Cycle Cost Workshop, Joint Services Data Exchange Group for Inertial Systems, Hosted by Tobyhanna Army Depot, Tobyhanna, Pennsylvania, USA October 1972, pp. 144–180.
530. W.B. Taylor, "Use of Life Cycle Costing in Acquiring Physical Assets", Long Range Planning, Vol. 14, No. 6, December 1981, pp. 32–43.
531. S.S. Thabit, "Study of the Technique of Life-Cycle Costing as Applied to the Appraisal of Oil Field Power Supply Systems", Engineering Costs of Production Ecnomics, Vol. 7, No. 4, March 1984, pp. 305–312.
532. S.S. Thabit, "Life Cycle Costings: A Decade of Progress", Chartered Mechanical Engineer, Vol. 30, No. 5, May 1983, pp. 46–49.
533. The American Institute of Architects, "Life Cycle Cost Analysis: A Guide for Architects", The American Institute of Architects, Washington, DC, 1977.
534. G. Thomann, J.A. Fulton, "Economic Analysis—Water Source Heat Pumps," ASHRAE Journal, Vol. 22, No. 9, September 1980, pp. 42–46.
535. U. Thomas, "Estimation of the Life Cycle Costs of Complex Technical Systems", Z Flugwiss Weltraumforsch (German), Vol. 7, No. 3, May–June 1983, pp. 213–220.
536. W. Thomas, "Procurement Costs Dictate Specification Tailoring", Proceedings of the Annual Reliability and Maintainability Symposium, 1979, pp. 384–387.
537. C.B. Thomsen, "An Overview of Cost Management", Architectural Record, Vol. 154, September 1973.

538. G.R. Thornber, "Impact of Maintainability on Life Cycle Costs", in Design to Cost and Life Cycle Cost, AGARD Conference Proceedings No. 289, May 1980, pp. 20-1 to 20-11. Available from National Technical Information Service (NTIS), Springfield, Virginia, U.S.A.
539. A. Thumann, "How Life Cycle Costing Cuts Yearly Operating Expenses", In Plant Engineers and Managers Guide to Energy Conservation, Van Nostrand Reinhold, Company, New York, 1977.
540. B. Tigerman, "Economy of Redundancy in Telecommunication Systems", Proceedings of the Annual Reliability and Maintainability Symposium, 1985, pp. 342–347.
541. E.A. Timsans, R.J. McNichols, S.L. Berry, "Availability Allocation Using a Family of Hyperbolic Cost Functions", IEEE Transactions on Reliability, Vol. R-24, No. 5, 1975, pp. 333–335.
542. F.S. Timson, "Practicality of Life Cycle Cost Models as Aids to Acquisition Decision Making: Confidence in Estimates", Rand Report No. WN8237-ARPA, The Rand Corporation, Santa Monica, California, 1978.
543. E.F. Toohey, A.B. Calvo, "Cost Analyses for Avionics Acquisition", Proceedings of the Annual Reliability and Maintainability Symposium, 1980, pp. 85–90.
544. E.P. Trott, "Maintainability and Reliability Cost Effectiveness", Annals of Reliability and Maintainability, Vol. 4, 1965, pp. 219–228.
545. I. Turiel, H. Estrada, and M. Levine, "Life-cycle Cost Analysis of Major Appliances", Energy, Vol. 6, No. 9, 1981, pp. 945–970.
546. D.E. Tuttle, M.N. Shwartz, "Lower Avionic Temperature Lower Life Cycle Cost", Proceedings of the Annual Reliability and Maintainability Symposium, 1979, pp. 332–337.
547. S. Tzemos, D. Dippold, "Stochastic Cost Estimating in Repository Life-Cycle Cost Analysis", Proceedings of the Annual Meeting of the Nuclear Materials Management, Vol. 15, New Orleans, LA, USA, June 22–25, 1986, pp. 279–290.
548. R. Urien, "Some Thoughts About the Economic Justification of Life Cycle Costing Formulae", Industrialization Forum, Vol. 6, No. 3–4, 1975, pp. 53–62.
549. US Comptroller General, "Study of Health Facilities Construction Costs", Washington, DC, Government Printing Office, 1972.
550. US Department of the Air Force, Air Force Systems Command, "Life Cycle Costing Special Provisions for UHF Modernization Radio Program ARC (XXX)", Revision 15, Aeronautical Systems Division, Wright Patterson AFB, Ohio, March 1974.
551. US Department of the Air Force, Air Force Systems Command and Air Force Logistics Command, "Review of the Application of Life-Cycle Costing to the A-X/A-10 Program (1970–1973)", Joint AFSC/AFLC

Commanders Working Group on Life Cycle Cost, Wright Patterson AFB, Ohio, October 1973.

552. US Department of the Air Force, Air Force Systems Command and Air Force Logistics Command, "Life Cycle Cost Reference Notebook", Compiled by the Joint AFSC/AFLC Commanders Working Group on Life Cycle Cost, Wright Patterson AFB, Ohio, August 1973.
553. US Department of Army, Office of Chief of Engineers, "Engineering Economic Studies: Life Cycle Costing Instructions", Appendices B, C, 1972.
554. U.S. Navy, Aviation Supply Office, "Case Histories of Life Cycle Costing Procurement", O. Markowitz (ed.), US Navy, Aviation Supply Office, Philadelphia, April 1974.
555. W.R. Vanderzanden, "The Economics of Power Transformer Reliability—Reliability Improvement by Profit Incentive", Proceedings of the 1980 Reliability Conference for the Electric Power Industry, 1980, pp. 105–108.
556. T.C. Varley, "Planning Acquisition Research", Proceedings of the Annual Reliability and Maintainability Symposium, 1981, pp. 337–338.
557. J. Vernon, "Life Cycle Cost in Preliminary Engine Design", Society of Automotive Engineers (SAE), Aerospace Meeting, San Diego, California, November 27–30, 1978, 7 pages, SAE Paper No. 781032.
558. W.Q. Wagner, "Engine Design Decisions Impact Aircraft Life Cycle Costs", AIAA Paper 77-916, July 1977, 9 pages. Available from the American Institute of Astronautics and Aeronautics (AIAA), New York.
559. W.Q. Wagner, C.B. Pheatt, "Apsi Cost-Model and Method for Turbine Engine Design to Life Cycle Cost", American Institute of Aeronautics and Astronautics and Society of Automotive Engineers, 12th Propulsion Conference, Palo Alto, California, U.S.A., July 26–29, 1976, 7 pages. AIAA Paper 76-750, 76/07/00, 76 A 38247. Available from the American Institute of Aeronautics and Astronautics, New York.
560. G. Walker, "Logistics Forecasting for Achieving Low Life Cycle Costs", in the Application of Design to Cost and Life Cycle Cost to Aircraft Engines, AGARD Lecture Series No. 107, May 1980, pp. 6-1 to 6-25. Available from National Technical Information Service (NTIS), Springfield, Virginia, U.S.A.
561. G. Walker, "Life-Cycle Cost/System Effectiveness, Evaluation and Criteria", Report, The Boeing Aircraft Company, Seattle, USA, January 1974.
562. R.A. Walker, "Analyzing HVAC Life-Cycle Economics on a Calculator", Specifying Engineer, Vol. 40, No. 6, November 1978, pp. 133–137.
563. G.F. Walters, J.A. McCall, "Software Quality Metrics for Life Cycle Cost-Reduction", IEEE Transactions on Reliability, Vol. R-28, 1979, pp. 212–220.

564. R.C. Waters, "Cost Performance Expectations: Economists and Engineers", Engineering Economist, Vol. 23, No. 2, 1978, pp. 93–100.
565. I.A. Watson, "Cost Basis for Availability Target Setting", Terotechnica, Vol. 2, No. 2, May 1981, pp. 121–134.
566. F.A. Watts, "Evaluating and Selecting the Preferred Air-Breathing Weapon System", in the Application of Design to Cost and Life Cycle Cost to Aircraft Engines, AGARD Lecture Series No. 107, May 1980, pp. 9-1 to 9-16. Available from National Technical Information Service (NTIS), 5285 Port Royal Road, Springfield, Virginia 22161, U.S.A.
567. R. Weber, "Life Cycle Cost Impacts of Unsafe Designs", Proceedings of the Annual Reliability and Maintainability Symposium, 1977, pp. 120–123.
568. R. Weber, "A Practical Method of Optimizing System Life Cycle Costs Vs Availability", Microelectronics and Reliability, Vol. 17, No. 1, 1978, pp. 1–7.
569. S.F. Weber, "Economic Analysis of Alternative Envelope Designs for New Residences in the United States", Energy, Vol. 5, No. 1, January 1980, pp. 63–68.
570. K. Weir, "Using Cost Data to Optimize Reliability", Proceedings of the 22nd Annual Conference of the American Society for Quality Control. (ASQC), 1968, pp. 100–105.
571. W.T. Weir, "Decision Theory and Cost Modeling", Proceedings of the Annual Reliability and Maintainability Symposium, 1972, pp. 319–328.
572. R.B. Werden, "A Practical Method of Optimizing System Life Cycle Costs Vs Availability", Microelectronics and Reliability, Vol. 17, 1978, pp. 1–7.
573. G.E. White, P.F. Ostwald, "Life Cycle Costing", Management Accounting, January 1976, pp. 39–41.
574. K. Wickel, "Design to Cost and Systems' LCC", in Design to Cost and Life Cycle Cost, AGARD Conference Proceedings No. 289, May 1980, pp. 19-1 to 19-9. Available from National Technical Information Service (NTIS), Springfield, Virginia, U.S.A.
575. J.E. Williams, "Data Requirements for Life Cycle Costing", American Association of Cost Engineers (AACE), Bulletin, November/December, 1977.
576. J.E. Williams, "Life Cycle Costing: An Overview", Joint Conference of American Institute of Industrial Engineers (AIIA) and American Association of Cost Engineers, Washington, DC, October 1977.
577. J.E. Williams, "A Model for Predicting Life Cycle Building Costs: Case Study of Two Public Universities in Michigan", Doctoral dissertation, University of Michigan, Ann Arbor, 1973.
578. M.F. Wilson, "Design to Cost Application in Military Environment Means Changing Old Ways", Defense Managemennt Journal, Vol. 9, No. 3, July 1973, pp. 17–20.

579. G.W. Windolph, "Considerations in R&M Program Optimization", Proceedings of the Annual Reliability and Maintainability Symposium, 1979, pp. 1–6.
580. J. Winger, T. Bray, P. Halter, "Lower Health Costs from High Equipment Reliability", Proceedings of the Annual Reliability and Maintainability Symposium, 1979, pp. 203–211.
581. R.G. Wingerter, F.W. Glenn, "Measuring the Payback of QA Efforts", Proceedings of the Annual Reliability and Maintainability Symposium, 1986, pp. 466–468.
582. E.S. Winlund, et al., "Cost Effectiveness", Proceedings of the Aerospace Reliability and Maintainability Conference, 1964, pp. 23–44.
583. N.K. Womer, "The Treatment of Risk in Analysis: A Framework for Discussion", in Electronic Systems Effectiveness and Life Cycle Costing, Edited by J.K. Skwirzynski, Springer-Verlag, Berlin, 1983, pp. 621–631.
584. J.K.R. Wood, "A Method of Determining the Cost Effectiveness of Different Levels of Reliability and Maintainability for a Training Aircraft", in Operations Research and Reliability, Edited by D. Grouchko, Gordon and Breach Science Publishers, New York, 1971, pp. 145–169.
585. H.W. Wynholds, J.P. Skratt, "Life Cycle Cost Analysis Program Users Guide", Report No. EM-2081-6, The Analytic Science, Reading, Massachusetts, 1979.
586. H.W. Wynholds, J.P. Skratt, "Weapon System Parametric Life Cycle Cost Analysis", Proceedings of the Annual Reliability and Maintainability Symposium, 1977, pp. 303–309.
587. H.W. Wynholds, J.P. Skratt, "Life Cycle Costing Saves Millions When Designing Industrial Plants", Airconditioning, Heating and Refrigeration News, No. 139, 1976, pp. 31–32.
588. H.W. Wynholds, J.P. Skratt, "Life Cycle Costing is Key to Conservation", Engineering News Record, Vol. 195, 1975, pp. 48–49.
589. R.R. Yanuck, "Effects of Fuel Escalation on Life Cycle Costing", Energy Engineering, August/September 1980, pp. 15–18.
590. T.R. Yezerski, M.L. Miller, "Economics of Power Plant RAM—Pennsylvania Power and Light Co. ECORAM Capability Described", Proceedings of the 13th INTER-RAM Conference for the Electric Power Industry, 1986, pp. 365–372.
591. W. Zabka, "Design-to-Cost, Design-to-Life-Cycle-Cost", Report No. UT-61-85. Lecture Given at the University of Dhahran, Saudi Arabia, April 1985. Available from Messerschmitt-Bolkow-Blohm Gmbh, Unternehmensbereich Apporate Dien Stieistungswerk Ottobrunn, Technisch-Wissenschafflische Information, Post fach 801220, 8000 Munchen 80, Federal Republic of Germany.
592. F. H. Zalud, and J. Lanc, "Automobile Reliability—A Key to Lower Overall Transport Costs", Proceedings of the XIVth International

Automobile Technical Congress of FISITA, London, England, 25–30 June, 1972, pp. 6/25–6/32. Available from the Institution of Mechanical Engineers, 1 Birdcage Walk, Westminster, London SW1H 9JJ, England.

A.3 Reliability Improvement Warranty

1. J.R. Anderson, "Warranties—The Easy Way Out", Proceedings of the Annual Reliability and Maintainability Symposium, 1979, pp. 413–415.
2. H. Balaban, M.A. Meth, "Contractor Risk Associated with Reliability Improvement Warranty", Proceedings of the Annual Reliability Improvement Warranty", Proceedings of the Annual Reliability and Maintainability Symposium, 1978, pp. 123–129.
3. H. Balaban, D. Cuppett, G. Harrison, "The F-16 RIW Porgram", Proceedings of the Annual Reliability and Maintainability Symposium, 1979, pp. 79–82.
4. H. Balaban, B. Retterer, "The Use of Warranties for Defense Avionics Procurement", Proceedings of the Annual Reliability and Maintainability Symposium, 1974, pp. 363–368.
5. H. Bayer, R.N. Speir, "Long Term Commercial Warranty", Proceedings of the Annual Reliability and Maintainability Symposium, 1978, pp. 50–54.
6. W.R. Blischke, "Calculation of the Cost of Warranty Policies as a Function of Estimated Life Distributions", Naval Research Logistics Quarterly, Vol. 22, 1975, pp. 681–696.
7. W.J. Bonner, "Warranty Contract Impact on Product Liability", Proceedings of the Annual Reliability and Maintainability Symposium, 1977, pp. 261–263.
8. W.J. Bonner, " A Contractor View of Warranty Contracting", Proceedings of the Annual Reliability and Maintainability Symposium, 1976, pp. 351–356.
9. T.N. Bowers, "Reliability Guarantee Versus Cost of Capital", Proceedings of the Annual Symposium on Reliability, 1969, pp. 157–163.
10. T.M. Cappels, "Full Cycle Corrective Action (FCCA) for Improved Warranty Service", Proceedings of the Annual Reliability and Maintainability Symposium, 1983, pp. 20–23.
11. P.O. Chelson, "Can we Expect ECPs Under RIW?", Proceedings of the Annual Reliability and Maintainability Symposium, 1978, pp. 204–209.
12. R.C. Day, L.E. McIntyre, "RIW Data Collection and Reporting Method", Proceedings of the Annual Reliability and Maintainability Symposium, 1978, pp. 66–72.
13. E.F. Dertinger, "Status of Reliability Requirements in Government Contracts", Proceedings of the Annual Reliability and Quality Control Symposium, 1965, pp. 125–135.

14. B.S. Dhillon, "Systems Reliability", Maintainability and Management, Petrocelli Books Inc., New York, 1983, pp. 155–164.
15. J.A. DiPaolantonio, R.H. Kohaya, R.T. Miyamoto, B.E. Waller, "Case Report: R&M Support of CLS Warranties", Proceedings of the Annual Reliability and Maintainability Symposium, 1986, pp. 457–459.
16. S.G. Dizek, "Automated Decision Support for Warranty Selection", Proceedings of the Annual Reliability and Maintainability Symposium, 1986, pp. 460–465.
17. N.G. Fleig, "The United States Air Force Product Performance Agreement Center", Proceedings of the Annual Reliability and Maintainability Symposium, 1986, pp. 449–453.
18. W.W. Flottman, M.R. Worstell, "Mutual Development, Application and Control of Suppliers Warranties", Proceedings of the Annual Reliability and Maintainability Symposium, 1977, pp. 213–221.
19. R.K. Gates, R.S. Bicknell, J.E. Bortz, "Quantitative Models Used in the RIW Decision Process", Proceedings of the Annual Reliability and Maintainability Symposium, 1977, pp. 229–236.
20. R.M. Genet, "Government Depot Maintenance Warranties", Proceedings of the Annual Reliability and Maintainability Symposium, 1976, pp. 363–365.
21. A.J. Glaser, "A Data Information System for RIW Contracts", Proceedings of the Annual Reliability and Maintainability Symposium, 1981, pp. 139–143.
22. A.F. Grimm, "A Method for Predicting Warranty Costs", Proceedings of the American Society for Quality Control Annual Conference, 1978, pp. 205–210.
23. P.B. Guglielmoni, "An Approach to Selecting Warranties/Incentives", Proceedings of the Annual Reliability and Maintainability Symposium, 1986, pp. 166–170.
24. C.A. Hardy, R.J. Allen, "Reliability Improvement Warranty Techniques and Applications", Proceedings of the Annual Reliability and Maintainability Symposium, 1977, pp. 222–228.
25. A.J. Hauter, C.W. Strempke, "Tacan RIW Program", Proceedings of the Annual Reliability and Maintainability Symposium, 1978, pp. 62–65.
26. A.H. Hevesh, "Cost of Reliability Improvement", Proceedings of the Annual Symposium on Reliability, 1969, pp. 54–61.
27. R. M. Jacobs, and L. A. Hoffman, "Contracting for Liability Relief", Proceedings of the Annual Reliability and Maintainability Symposium, 1984, pp. 67–71.
28. C.R. Knight, "Warranties as a Life Cycle Cost Management Tool", Defense Management Journal, Vol. 12, No. 1, January 1976, pp. 23–28.

29. R. Kowalski, R. White, "Reliability Improvement Warranty (RIW) and the Army Lightweight Doppler Navigation Systems (LDNS)", Proceedings of the Annual Reliability and Maintainability Symposium, 1977, pp. 237–241.
30. D.H. Kruvand, "Army Aviation Warranty Concepts", Proceedings of the Annual Reliability and Maintainability Symposium, 1987, pp. 392–394.
31. J.D. Lunsford, J.H. Berk, D.E. Nixon, "Warranting Conventional Munitions", Proceedings of the Annual Reliability and Maintainability Symposium, 1986, pp. 443–448.
32. O. Markowitz, "How to Build a Long Term Warranty Strategy", Proceedings of the Reliability Conference for the Electric Power Industry, 1978, pp. 99–103.
33. O. Markowitz, "A New Approach Long Range Fixed Price Warranty within Operational Environments—for Buyer/User", Annals of Reliability and Maintainability, 1971, pp. 252–257.
34. O. Markowitz, "Continuous Reliability Improvement Assurance within Operations and Logistics", Proceedings of the 1977 Reliability Conference for the Electric Power Industry, 1977, pp. 100–105.
35. O. Markowitz, "Aviation Supply Office FFW/RIW Case History #2, Abex Pump", Proceedings of the Annual Reliability and Maintainability Symposium, 1976, pp. 357–362.
36. C.W. Marshall, "Design Trade-Offs in Availability Warranties", Proceedings of the Annual Reliability and Maintainability Symposium, 1981, pp. 95–100.
37. E.P. McGuire, "Industrial Product Warranties: Policies and Practices", A Research Report from the Conference Board, Report No. 800, 1980. Available from the Conference Board, Inc., 845 Third Avenue, New York, NY 10022, U.S.A.
38. Massachusetts Institute of Technology (MIT), "Appliance Warranties and Service Contracts: Description, Pricing and Legal Aspect", Report No. CPA-78-14, Vol. II, MIT, Massachusetts, December 1978.
39. Massachusetts Institute of Technology (MIT), "Appliance Warranties and Service Contracts: Consumer Experience", Report No. CPA-78-14, Vol. III, MIT, Massachusetts, December 1978.
40. Massachusetts Institute of Technology (MIT), "Policy Alternatives for the Problem of Product Failure", Report No. CPA-78-14, Vol. 1, MIT, Massachusetts, December 1978.
41. R.A. Mlinarchik, "Assuring Army User Satisfaction Through Warranties", Proceedings of the Annual Reliability and Maintainability Symposium, 1980, pp. 307–313.
42. P.O. Nerber, "Warranty Cost Estimates for Avionic Subsystems", Proceedings of the Annual Symposium on Reliability, 1969, pp. 270–275.

43. D.G. Newman, L.D. Nesbitt, "USAF Experience with RIW", Proceedings of the Annual Reliability and Maintainability Symposium, 1978, pp. 55–61.
44. K.S. Park, "Optimal Use of Product Warranties", IEEE Transactions on Reliability, Vol. R-34, No. 5, December, 1985, pp. 519–521.
45. E.A. Polgar, J.H. Yuch, "A Model for Determination of Incentive Fee", Proceedings of the 11th National Symposium on Reliability and Quality Control, 1965, pp. 115–124.
46. B.L. Retterer, "Considerations for Effective Warranty Application", Proceedings of the Annual Reliability and Maintainability Symposium, 1976, pp. 346–350.
47. M.R. Seldon, "Estimating the Cost of R and M Changes", Proceedings of the Annual Reliability and Maintainability Symposium, 1980, pp. 79–84.
48. D.K. Shelton, R.G. Paxman, "A Reliability Warranty Concept for the FMS Environment", Proceedings of the Annual Reliability and Maintainability Symposium, 1982, pp. 34–39.
49. R.R. Shorey, "Factors in Balancing Government and Contractor Risk with Warranties", Proceedings of the Annual Reliability and Maintainability Symposium, 1976, pp. 366–368.
50. R.M. Springer, Jr., "RIW with Cost Sharing", Proceedings of the Annual Reliability and Maintainability Symposium, 1977, pp. 391–395.
51. M.U. Thomas, "Optimum Warranty Policies for Nonrepairable Items", IEEE Transactions on Reliability, Vol. R-32, No. 3, August 1983, pp. 282–288.
52. C.D. Weimer, "Contractor Initiatives for R&M/Cost Improvement, Proceedings of the Annual Reliability and Maintainability Symposium, 1978, pp. 243–250.
53. R.P. Wilcox, "Failure Free Warranty—5 Year Results", Proceedings of the IEEE 1974 National Aerospace and Electronics Conference (NAECON'74), Dayton, Ohio, May 1974, pp. 564–571.
54. A.R. Yuspeh, "Legislation on Weapon System Warranties", Proceedings of the Annual Reliability and Maintainability Symposium, 1986, pp. 438–442.

A.4 Software Costs

1. B.S. Dhillon, Reliability in Computer System Design, Ablex Publishing Corporation, Norwood, New Jersey 07648, USA, 1987.
2. D.V. Ferens, R.L. Harris, "Avionics Computer Software Operation and Support Cost Estimation", Proceedings of the IEEE 1979 National Aerospace and Electronics Conference (NAECON 1979), Dayton Convention Center, Dayton, May 15–17, 1979, pp. 296–300.

3. M.F. Finfer, R.K. Mish, "Software Acquisition Management Guidebook: Software Cost Estimation and Measurement", Report No. ESD-TR-78-140, March 1978, 114 pp. Available from National Technical Information Service, Springfield, VA 22161, U.S.A.
4. M. Friedman, "Modeling the Penalty Costs of Software Failure", Proceedings of the Annual Reliability and Maintainability Symposium, 1987, pp. 359–363.
5. A.L. Goel, J.B. Soenjoto, "Hardware-Software Availability: A Cost Based Trade-off Study", Proceedings of the Annual Reliability and Maintainability Symposium, 1983, pp. 303–309.
6. W.D. Goeller, "The Cost of Software Quality Assurance", American Society for Quality Control (ASQC) Quality Congress Transactions, San Francisco, U.S.A., 1981, pp. 684–689.
7. C.P. Hollocker, "Finding the Cost of Software Quality", IEEE Transactions on Engineering Management, Vol. EM-33, No. 4, November 1986, pp. 223–228.
8. T.G. James, Jr., "Software Cost Estimating Methodology", Proceedings of the IEEE 1977 National Aerospace and Electronics Conference (NAECON'77), Dayton Convention Center, Dayton, May 17–19, 1977, pp. 22–28.
9. L.H. Putnam, A. Fitzsimmons, "Estimating Software Costs, Part I, In Software Cost Estimating and Life-cycle Control, Edited by L.H. Putnam, Computer Society Press, IEEE Computer Society, New York, 1980, pp. 312–315.
10. L.H. Putnam, "Example of an Early Sizing, Cost and Schedule Estimate for an Application Software System", in Software Cost Estimating and Life Cycle Control, Computer Society Press, IEEE Computer Society, New York, 1980, pp. 102–127.
11. L.H. Putnam, A. Fitzsimmons, "Estimating Software Costs", Part III, in Software Cost Estimating and Life-cycle Control", Computer Society Press, IEEE Computer Society, New York, 1980, pp. 321–323.
12. L.H. Putnam, A. Fitzsimmons, "Estimating Software Costs", Part II, In Software Cost Estimating and Life-cycle Control, Edited by L.H. Putnam, Computer Society Press, IEEE Computer Society, New York, 1980, pp. 316–320.
13. L.H. Putnam, R.W. Wolverton, "Tutorial: Quantitative Management, Software Cost Estimating", COMPSAC 77, The IEEE Computer Society First International Computer Software & Applications Conference, Chicago, November 8–11, 1977.
14. V. Schneider, "Prediction of Software Effort and Project Duration—Four New Formulas", ACM SIGPLAN Notices, Volume 13, No. 6, June 1978, pp. 49–59.
15. M. Stanley, "Software Cost Estimating", Royal Signals & Radar Establishment (RSRE) Memorandum No. 3472, 13th May 1982, London, U.K.

16. R.W. Wolverton, The Cost of Developing Large-Scale Software", IEEE Transactions on Computers, June 1974, pp. 282–303.
17. R.W. Wolverton, "Software Costing", in Handbook of Software Engineering, Edited by C. R. Vick, C. V. Ramamoorthy, Van Nostrand Reinhold Company, New York, 1984, pp. 469–493.

A.5 Manufacturing Costs

1. J.W. Hackney, "Estimate Production Costs Quickly", in Modern Cost-Engineering Techniques, Edited by H.O. Popper and the Staff of Chemical Engineering, McGraw-Hill Book Company, Inc., New York, 1970, pp. 245–250.
2. I. Ham, "Group Technology", in Handbook of Industrial Engineering, Edited by G. Salvendy, John Wiley & Sons, Inc., New York, 1982, pp. 7.8.1.–7.8.19.
3. E.M. Malstrom (Editor), Manufacturing Cost Engineering Handbook, Marcel Dekker, Inc., New York, 1984.
4. M. Phister, Jr., "Analyzing Computer Technology Costs—Part I: Development and Manufacturing", Computer Design, September 1978, pp. 91–98.
5. The Cost Estimating Committee, "Cost Estimating", in Manufacturing Planning and Estimating Handbook, Edited by Frank W. Wilson, and E.P.D. Harvey, McGraw-Hill Book Company, New York, 1963, Chapter 3.

A.6 Quality Costs

1. American Society for Quality Control (ASQC), "Guidelines for Reducing Quality Costs", Prepared by Quality Cost Technical Committee, American Society for Quality Control, Milwaukee, Wisconsin, 1977.
2. Americn Society for Quality Control, "Quality Costs—What & How", Second Edition, Prepared by the Quality Cost-Cost Effectiveness Committee, American Society for Quality Control, Milwaukee, 1971.
3. C.A. Aubrey II, D.A. Zimbler, " A Banking Quality Cost Model, its Uses and Results", American Society for Quality Control (ASQC) Quality Congress Transactions, Detroit, 1982, pp. 195–201.
4. S.K. Bhuyan, "Cost of Quality as a Customer Perception", American Society for Quality Control (ASQC) Quality Congress Transactions, Detroit, 1982, pp. 459–464.
5. J.D. Breeze, "Quality Costs can be Sold", American Society for Quality Control (ASQC) Technical Conference Transactions, Atlanta, 1980, pp. 795–801.

6. J.D. Breeze, J.R. Farrell, "Quality Costs can be Sold: Part II", American Society for Quality Control (ASQC) Quality Congress Transactions, San Francisco, 1981, pp. 737–741.
7. C.W. Brewer, " "Q" Costs/Results—The Petroleum Industry", American Society for Quality Control (ASQC) Quality Congress Transactions, Detroit, 1982, pp. 872–883.
8. F.X. Brown, "Quality Costs and Strategic Planning", American Society for Quality Control (ASQC) Technical Conference Transactions, Atlanta, 1980, pp. 155–159.
9. BS 6143, "Determination and Use of Quality-Related Costs", British Standards Institution, London, 1981.
10. W.O. Cabral, "Quality Cost Myopia", American Society for Quality Control (ASQC) Quality Congress Transactions, Boston, 1983, pp. 594–598.
11. J. Campanella, F.J. Corcoran, "Principles of Quality Costs", American Society for Quality Control (ASQC) Quality Congress Transactions, Detroit, 1982, pp. 124–137.
12. J. Campanella, "The Fairchild Republic Company Quality Cost Program", American Society for Quality Control (ASQC) Technical Conference Transactions, Houston, 1979, pp. 584–593.
13. J.K. Carson, "Quality Costing—A Practical Approach", International Journal of Quality & Reliability Management, Vol. 3, No. 1, 1986, pp. 54–63.
14. B. Cox, "Interface of Quality Costing and Technology", First International Congress on Technology, London, U.K., May 1979, pp. 2B.1.1–2B.1.6.
15. E.W. Dawes, "Reducing Appraisal Costs", American Society for Quality Control (ASQC) Technical Conference Transactions, Chicago, 1978, pp. 650–654.
16. J. Demetriou, "Cost of Quality System—A Management Tool", American Society for Quality Control (ASQC) Quality Congress Transactions, Boston, 1983, pp. 378–380.
17. J. Demetriou, "Quality Costs—Pay", American Society for Quality Control (ASQC) Quality Congress Transactions, Detroit, 1982, pp. 860–865.
18. B.S. Dhillon, Quality Control, Reliability, and Engineering Design, Marcel Dekker, Inc., New York, U.S.A., 1985.
19. R.K. Dobbins, "Extending Effectiveness of Quality Cost Programs", American Society for Quality Control (ASQC) Technical Conference Transactions, Chicago, 1978, pp. 200–204.
20. R.K. Dobbins, "Your Quality Posture—A Profit/Survival Challenge", American Society for Quality Control (ASQC) Quality Congress Transactions, San Francisco, 1981, pp. 485–490.
21. L.J. Esterby, "Quality Cost Analysis: A Productivity Measure", American Society for Quality Control (ASQC) Quality Congress Transactions, San Francisco, 1981, pp. 432–439.

22. L.J. Esterby, "Measuring Quality Costs by Work Sampling", American Society for Quality Control (ASQC) Quality Congress Transactions, Detroit, 1982, pp. 138–144.
23. A.O. Gunneson, "How to Effectively Implement a Quality Cost System", American Society for Quality Control (ASQC) Quality Congress Transactions, San Francisco, 1981, pp. 99–106.
24. H.J. Harrington, Poor-Quality Cost, Marcel Dekker, Inc., New York, 1987.
25. H.J. Harrington, "Quality Cost—A Key to Productivity", American Society for Quality Control (ASQC) Quality Congress Transactions, San Francisco, 1981, pp. 420–431.
26. S. Houchheiser, R.C. Rhodes, "Quality Costs for Environmental Monitoring Systems", American Society for Quality Control (ASQC) Technical Conference Transactions, Philadelphia, 1977, pp. 328–334.
27. D.L. Kimble, "Bayesian Cost Analysis of Rectifying Rejected Lots", American Society for Quality Control (ASQC) Technical Conference Transactions, Houston, 1979, pp. 756–759.
28. R.C. Kroeger, "Quality Costs—A New Perspective", American Society for Quality Control (ASQC) Technical Conference Transactions, Houston, 1979, pp. 575–583.
29. W.J. Latzko, "Minimizing the Cost of Inspection", American Society for Quality Control (ASQC) Quality Congress Transactions, Detroit, 1982, pp. 485–490.
30. D.D. Lee, "Minimized Cost Sampling Technique", American Society for Quality Control (ASQC) Quality Congress Transactions, San Francisco, 1981, pp. 958–963.
31. B.S. Liebesman, "Selection of MIL-STD-105D Plans Based on Costs", American Society for Quality Control (ASQC) Quality Congress Transactions, San Francisco, 1981, pp. 475–484.
32. W.N. Moore, "The Philosophy and Usefulness of Quality Costs", American Society for Quality Control (ASQC) Technical Conference Transactions, Chicago, 1978, pp. 644–649.
33. NEDC Task Force on Quality and Standards, "Quality and Value for Money", National Economic Development Council (NEDC), London, May 1985.
34. J.J. Plunkett, B.G. Dale, "A Review of the Literature on Quality-Related Costs", International Journal of Quality & Reliability Management, Vol. 4, No. 1. 1987, pp. 40–52.
35. F. Scanlon, "Cost Reduction Through Quality Management", American Society for Quality Control (ASQC) Technical Conference Transactions, Atlanta, 1980, pp. 270–273.
36. H.E. Schock, Jr., "Societal Quantity and Regulatory Costs", American Society for Quality Control (ASQC) Technical Conference Transactions, Atlanta, 1980, pp. 656–661.

37. D. Shainin, P.D. Krensky, E.W. Dawes, "Can Quality Cost Principles be Applied to Product Liability?", American Society for Quality Control (ASQC) Technical Conference Transactions, Houston, 1979, pp. 295–296.
38. W. Siff, "Quality Costs in the Process Industries", American Society for Quality Control (ASQC) Quality Congress Transactions, Boston, 1983, pp. 207–210.
39. D.S. Sink, J.B. Keats, "Using Quality Costs in Productivity Measurement", American Society for Quality Control (ASQC) Quality Congress Transactions, Boston, 1983, pp. 211–216.
40. E.T. Szymanski, "Overcoming Regulatory Constraints in Quality Costs", American Society for Quality Control (ASQC) Quality Congress Transactions, Detroit, 1982, pp. 8–12.
41. J.J. Wayne, A.F. Grimm, "Quality Costs: We Know Where We're Going! Do You?", American Society for Quality Control (ASQC) Quality Congress Transactions, Boston, 1983, pp. 457–462.
42. F.B. Whitehall, "Review of Problems with a Quality Cost System", International Journal Quality & Reliability Management, Volume 3, No. 3, 1986, pp. 43–58.
43. R.J. Williams, "Guide for Reducing Quality Costs", American Society for Quality Control (ASQC) Quality Congress Transactions, Detroit, 1982, pp. 360–366.
44. M.F. Wilson, M.L. Woodruff, "Economic Benefits of Part Quality Knowledge", Proceedings of the Annual Reliability and Maintainability Symposium, 1985, pp. 17–19.
45. W.O. Winchell, "The Hidden Aspect of Vendor Quality Cost", American Society for Quality Control (ASQC) Technical Conference Transactions, Philadelphia, 1977, pp. 236–239.
46. W.O. Winchell, "Guide for Managing Vendor Quality Costs", American Society for Quality Control (ASQC) Quality Congress Transactions, San Francisco, 1981, pp. 690–694.
47. W.O. Winchell, "Guide for Managing Vendor Quality Costs", American Society for Quality Control (ASQC) Technical Conference Transactions, Atlanta, 1980, pp. 483–487.
48. W.O. Winchell, "Guide for Managing Vendor Quality Costs", American Society for Quality Control (ASQC) Quality Congress Transactions, Detroit, 1982, pp. 367–372.
49. W.O. Winchell, "Reducing Failure Costs and Measuring Improvement", American Society for Quality Control (ASQC) Technical Conference Transactions, Chicago, 1978, pp. 655–657.
50. A.H. Zaludova, F.H. Zalud, "Quality Optimization via Total Quality Costs", American Society for Quality Control (ASQC) Quality Congress Transactions, San Francisco, 1981, pp. 729–736.
51. J.F. Zerfas, L. Arrington, "Guide for Reducing Quality Costs", American

Society for Quality Control (ASQC) Technical Conference Transactions, Atlanta, 1980, pp. 56–59.
52. W.C. Noz, Jr., B.F. Redding, P.A. Ware, "The Quality Managers Job: Optimize Costs," American Society for Quality Control (ASQC) Quality Congress Transactions, Boston, 1983, pp. 301–306.

A.7 Cost Estimation

1. N.G. Bach, "How to get More Accurate Plant Cost Estimates", in Cost Engineering in the Process Industries, Edited by C.H. Chilton and the Staff of Chemical Engineering, McGraw-Hill Book Company, Inc., New York, 1960, pp. 15–19.
2. J. Bollman, H. Lankenau, "Estimation of Relative Total Cost for Aircraft Systems", in Design to Cost and Life Cycle Cost, AGARD Conference Proceedings No. 289, May 1980, pp. 21-1 to 21-9. Available from National Technical Information Service (NTIS), Springfield, Virginia, USA.
3. H.E. Boren, "Cost-Estimating Relationships for Aircraft Airframes", Report No. R-761-PR, The Rand Corporation, Santa Monica, California, USA, February 1972.
4. H.E. Boren, "DAPCA: A Computer Program for Determining Aircraft Development and Production Costs", Report No. RM-5221-PR, The Rand Corporation, Santa Monica, California, USA, February 1967.
5. H.E. Boren, "Individual Weapon System Computer Cost Model", RM-4165-PR, The Rand Corporation, Santa Monica, California, USA, July 1964.
6. J.H. Black, "Operating-Cost Estimation", in Cost and Optimization Engineering, Edited by F.C. Jelen and J.H. Black, McGraw-Hill Book Company, Inc., New York, 1983, pp. 382–406.
7. R.M. Braca and J. Happel, "New Cost Data bring Economic Pipe Sizing Up to Date", in Cost Engineering in the Process Industries, Edited by C.H. Chilton and the Staff of Chemical Engineering, McGraw-Hill Book Company, Inc., New York, 1960, pp. 210–217.
8. R. Brown, "Economic Evaluation of Future Equipment Needs", Chemical Engineering, January 17, 1977, pp. 125–127.
9. R.L. Buchanan, "How to Avoid Common Pitfalls in Making Cost Estimates", Chemical Engineering, October 25, 1965, pp. 159–162.
10. J.P. Bullis, "Engineering Cost Control", Cost Engineering, Vol. 29, No. 4, April 1987, pp. 16–18.
11. H.G. Campbell, et al., "An Introduction to Equipment Cost Estimating", Report No. RM-6103-SA, The Rand Corporation, Santa Monica, California, December 1969.
12. K.E. Case, "On the Consideration of Variability in Cost Estimating", IEEE Transactions on Engineering Management, Vol. EM-19, No. 4, November 1972, pp. 114–118.

13. C.H. Chilton, "Cost Estimating Simplified", in Cost Engineering in the Process Industries, Edited by C.H. Chilton and the Staff of Chemical Engineering, McGraw-Hill Book Company, Inc., New York, 1960, pp. 28–31.
14. R.E. Clapp, "MISCOM: An Individual Missile Weapon System Computer Cost Model", Report No. RM-5142-1-PR, The Rand Corporation, Santa Monica, California, February 1967, "Addendum to RM-5142-1-PR: Modifications to the MISCOM Computer Cost Model RM-5142-1-PR (Addendum)", The Rand Corporation, Santa Monica, California, March 1967.
15. C.A. Collier, W.B. Ledbetter, Engineering Cost Analysis, Harper & Row, Publishers, New York, 1982.
16. A.B. Corripio, K.S. Chrien, L.B. Evans, "Estimate Costs of Heat Exchangers and Storage Tanks via Correlations", Chemical Engineering, January 25, 1982, pp. 125–127.
17. A.B. Corripio, K.S. Chrien, L.B. Evans, "Estimate Costs of Centrifugal Pumps and Electric Motors", Chemical Engineering, February 22, 1982, pp. 115–118.
18. E.A. Criner, Successful Cost Reduction Programs for Engineers and Managers, Van Nostrand Reinhold Company, New York, 1984.
19. M.B. Desai, "Preliminary Cost Estimating of Process Plants", Chemical Engineering, July 27, 1981, pp. 65–70.
20. P.F. Dieneman, "Estimating Cost Uncertainty Using Monte Carlo Techniques", Report No. RM-4854-PR, The Rand Corporation, Santa Monica, California, January 1966.
21. G.E. Dieter, "Cost Evaluation", in Engineering Design: A Materials and Processing Approach, McGraw-Hill Book Company, Inc., U.S.A., 1983, pp. 324–366.
22. E.N. Dodson, S.F. Fornish, R.R. Lieberman, W.E. Waller, "Cost Analysis of Avionics Equipment", Report No. AD781132, February 1974, 132 pp. Available from National Technical Information Service, U.S. Department of Commerce, 5285 Port Royal Road, Springfield, VA 22151, U.S.A.
23. D.L. Doty, P.J. Nelson and K.R. Stewart, "Software Cost Estimation Study, Volume II: Guidelines for Improved Software Cost Estimating", Report No. RADC-TR-77-220, Volume II, August 1977, 145 pp., Rome Air Development Center, Griffis AFB, NY 13411, U.S.A.
24. C.E. Dryden, R.H. Furlow, "Chemical Engineering Costs: A Condensed Handbook of Chemical Plant Equipment and Process Costs for Cost Estimating Purposes", Engineering Experiment Station, Ohio State University, Columbus, Ohio, USA, 1973, 73 pp.
25. W.H. Edgerton, "Understanding Cost Trends, Prelude to Estimating", Architectural Record, Vol. 145, May 1969, pp. 91.
26. W.B. Elsner, "Dimensioning Trunk Groups for Digital Networks", The

Bell System Technical Journal, Vol. 59, No. 7, September 1980, pp. 1123–1138.

27. A. Foord, "Investment Appraisal: An Academic Approach for Engineering Studies", AACE (American Association of Cost Engineers) Bulletin, October 1972, pp. 153–155.
28. L. Fried, "Estimating the Cost of System Implementation—Part II", Data Processing Magazine, April 1969, pp. 24–28.
29. J.T. Gallagher, "Rapid Estimation of Plant Costs", in Modern Cost-Engineering Techniques, Edited by H. Popper, et al., McGraw-Hill Book Company, New York, 1970, pp. 3–10.
30. J.T. Gallagher, "Analyzing Field Construction Costs". in Modern Cost Engineering Techniques, Edited by H. Popper and the Staff of Chemical Engineering, McGraw-Hill Book Company, New York, 1970, pp. 226–231.
31. S. Gets, "Techniques for Estimating Logistics Support and Operations Costs of Naval Airborne Weapon Systems", Report No. NADC-SD-6925, Naval Air Development Center, USA, April 1969.
32. R.W. Grubbstrom, H.H. Hinterhuber (Editors), "Special Issue of Engineering Costs and Production Economics", Proceedings of the Second International Working Seminar on Production Economics, Lans/Innsbruck, Austria, February 16–20, 1981, Published by Elsevier, Amsterdam, 1982, 276 pp.
33. J.W. Hackney, "Estimating Methods for Process Industry Capital Costs", in Modern Cost-Engineering Techniques, Edited by Popper and the Staff of Chemical Engineering, McGraw-Hill Book Company, Inc., New York, 1970, pp. 43–58.
34. J.W. Hackney, "Capital Cost Estimates for Process Industries", in Modern Cost-Engineering Techniques, Edited by H. Popper and the Staff of Chemical Engineering, McGraw-Hill Book Company, New York, 1970, pp. 25–42.
35. B.M. Hadfield, "Satellite-Systems Cost Estimation", IEEE Transactions on Communications, Vol. COM-22, No. 10, October 1974, pp. 1540–1547.
36. R.S. Hall, J. Matley, K.J. McNaughton, "Current Costs of Process Equipment", Chemical Engineering, April 5, 1982, pp. 80–116.
37. P.T. Heilig, "Navy Air-Launched Missile Operating and Support Cost Estimating Model", Administrative Sciences Corporation, Falls Church, Virginia, U.S., 1979.
38. J.H. Herd, J.N. Postak, W.E. Russell and K.R. Stewart, "Software Cost Estimation Study, Volume I: Study Results", Report No. RADC-TR-77-220-Volume I (of two), June 1977, 202 pp. Rome Air Development Center (ISIS), Griffis AFB, NY 13441, U.S.A.
39. J.T. Hinely, Jr., "Method for Estimating Operating and Maintenance Costs for C-5A Subsystems", Report, Lockheed Georgia Company, Marietta, Georgia, December 1970.

40. O. Igra, "Cost-Effectiveness of the Vortex-Augmented Wind Turbine", Energy, Vol. 4, 1979, pp. 119–130.
41. F.C. Jelen, J.H. Black (Editors), Cost and Optimization Engineering, McGraw-Hill Book Company, New York, 1983.
42. F.C. Jelen, "Major Cost Analysis Methods Yield", in Cost Engineering in the Process Industries, Edited by C.H. Chilton and the Staff of Chemical Engineering, McGraw-Hill Book Company, Inc., New York, 1960, pp. 376–378.
43. R.W. Jensen, "Macro-Level Software Development Cost Estimation Methodology", IEEE Transactions On Computers 1981, pp. 320–325.
44. S. Kasper, "Selecting Heat-Transfer Media by Cost Comparison", in Modern Cost-Engineering Techniques, Edited by H. Popper and the Staff of Chemical Engineering, McGraw-Hill Book Company, Inc., New York, 1970, pp. 468–471.
45. R.E. Kenyon, "Techniques for Estimating Weapon System Structural Costs", Report No. AD 747007, April 1972. Available from National Technical Information Service, US Department of Commerce, Springfield, VA 22151, U.S.A.
46. S.G. Kirkham, "Preparation and Application of Refined Lang Factor Costing Techniques", AACE (American Association of Cost Engineers) Bulletin, Vol. 14, No. 5, October 1972, pp. 137, 149, 157, 158.
47. I.V. Klumpar, S.T. Slavsky, "Updated Cost Factors: Commodity Materials", Chemical Engineering, August 19, 1985, pp. 76–77.
48. I.V. Klumpar, S.T. Slavsky, "Updated Cost Factors: Process Equipment", Chemical Engineering, July 22, 1985, pp. 73–75.
49. I.V. Klumpar, S.T. Slavsky, "Updated Cost Factors: Installation Labor", Chemical Engineering, September 16, 1985, pp. 85–87.
50. A.S. Korkotsides, K.T. Wallenius, "Parametric Cost Methodology", Proceedings of the Annual Reliability and Maintainability Symposium, 1981, pp. 346–352.
51. J.D. Kumana, "Cost Update on Specialty Heat Exchangers", Chemical Engineering, June 25, 1984, pp. 169–172.
52. V.L. Lambert, H.F. Sackett, "Research and Development Cost Estimation", IRE Transactions on Engineering Management, March 1959, pp. 8–12.
53. H.J. Lang, "Simplified Approach to Preliminary Cost Estimates", in Cost Engineering in the Process Industries, Edited by C.H. Chilton and the Staff of Chemical Engineering, McGraw-Hill Book Company, Inc., New York, 1960, pp. 12–14.
54. J.P. Large, "Development or Parametric Cost Models for Weapon Systems", Report No. AD A105179, Rand Corporation, Santa Monica, California, April 1981, 31 pages, NTIS Report No. ADA 105179. Available from the National Technical Information Service (NTIS), Springfield, Virginia.

55. P.R.G. Layard, D.W. Verry, "Cost Function for University Teaching and Research", Economic Journal, Vol. 85, No. 337, 1975, pp. 55–74.
56. R. Layard (Ed.), Cost-Benefit Analysis: Selected Readings, Penguin Books, Ltd., Harmondsworth, Middlesex, England, 1976, pp. 9–100.
57. G.S. Levenson, "Cost-Estimating Relationship for Aircraft Airframes", Report No. RM-4845-PR, The Rand Corporation, Santa Monica, California, USA, May 1966.
58. G.S. Levenson, et al., "Cost Estimating Relationships for Aircraft Airframes", Report No. R-761-PR, The Rand Corporation, Santa Monica, California, December 1971.
59. R.P. Lutz, "Discounted Cash Flow Techniques", in Handbook of Industrial Engineering, Edited by G. Salvendy, John Wiley & Sons, Inc., 1982, pp. 9.3.1.–9.3.25.
60. E.M. Malstrom (Editor), Manufacturing Cost Engineering Handbook, Marcel Dekker, Inc., New York, 1984.
61. G.E. Mapstone, "Evaluating Quantity Discounts", In Modern Cost-Engineering Techniques, Edited by H. Popper and the Staff of Chemical Engineering, McGraw-Hill Book Company, 1970, pp. 489–492.
62. A.M. McCosh, "Computerized Cost Classification System", Management Accounting, Vol. 53, April 1972, pp. 42–45.
63. J.D. McCullough, "Estimating System Costs", in Cost Effectiveness Analysis, Edited by T.A. Goldman, Praeger, Publishers, New York, 1968.
64. O. Mendel, "Estimating Engineering Cost", Chemical Engineering, December 9/23, 1985, pp. 117–118.
65. E.J. Mishan, Cost-Benefit Analysis: An Informal Introduction, George Allen & Unwin, Ltd., London, 1975.
66. A.L. Mular, "Mineral Processing Equipment Costs and Preliminary Capital Cost Estimations: A Handbook of Major Equipment Costs for Factored Capital Cost Estimations", Published by Canadian Institute of Mining and Metallurgy, Montreal, 1978, 166 pp.
67. A. Mulet, A.B. Corripio, L.B. Evans, "Estimate Costs of Pressure Vessels via Correlations", Chemical Engineering, October 5, 1981, pp. 145–150.
68. H.C. Nelson, O.P. Kharbanda, W.A. Janda, J.H. Black, "Capital Investment Cost Estimation", in Cost and Optimization Engineering, Edited by F.C. Jelen and J.H. Black, McGraw-Hill Book Company, New York, 1983, pp. 321–381.
69. D.B. Noreby, "Controlling and Estimating Labor Costs in the Construction Industry", Management Accounting, Vol. 51, May 1970, pp. 23–25.
70. G.T. O'Brien, "Estimating Design Costs", Machine Design, March 20, 1986, pp. 75–81.
71. E.O. Ohsol, "Estimating Marketing Costs", Chemical Engineering, May 3, 1971, pp. 116, 118, 120.

72. R.A. Oliver, "An Empirical Derivation of Cost Inputs for Quantitative Analysis", Journal of Systems Management, Vol. 21, October 1970, pp. 9–12.
73. P.F. Ostwald, "Cost Estimating", in Handbook of Industrial Engineering, Edited by G. Salvendy, John Wiley & Sons, Inc., New York, 1982, pp. 9.2.1.–9.2.28.
74. P.F. Ostwald, P.J. Toole, "IE's and Cost Estimating", Industrial Engineering, February 1978, pp. 40–43.
75. M. Phister, Jr., "Data Processing Technology and Economics", Second Edition, Santa Monica Publishing Company and Digital Press, Santa Monica, U.S.A., 1979.
76. A. Pikulik, H.E. Diaz, "Cost Estimating for Major Process Equipment", Chemical Engineering, October 10, 1977, pp. 106–122.
77. H. Popper and The Staff of Chemical Engineering (Editors), Modern Cost Engineering Techniques: An Economic-Analysis and Cost Estimation Manual, with Comprehensive Data on Plant and Equipment Costs in the Process Industries, McGraw-Hill, New York, 1970, 538 pp.
78. T.E. Powell, "A Review of Recent Developments in Project Evaluation", Chemical Engineering, November 11, 1985, pp. 187–194.
79. G.P. Purohit, "Estimating Costs of Shell-and-Tube Heat Exchangers", Chemical Engineering, August 22, 1983, pp. 56–57.
80. H. Ring, J. Robinson, "Method of Estimating Class IV Modification Costs for Fighter Aircraft", Report, Air Force Logistics Command, Wright-Patterson Airforce Base, Ohio, January 1970.
81. J.Y. Robinson, "New Data for Cost Estimating at the Conceptual Stage", American Institute of Architects (AIA) Journal, November 1974, pp. 33–37.
82. D.R. Sanchez, "Estimating Fixed-Wing Airframe Cost—Volume I", Report No. PRC R-547, Boeing Aerospace Company, Seattle, USA, February 1965.
83. S.C. Schuman, "How Plant Size Affects Unit Costs", in Cost Engineering in the Process Industries, Edited by C.H. Chilton and the Staff of Chemical Engineering, McGraw-Hill Book Company, Inc., New York, 1960, pp. 307–310.
84. S. Sobel, "A Computerized Technique to Express Uncertainty in Advanced System Cost Estimates", Report, The MITRE Corporation, Bedford, Massachusetts, 1975.
85. D.N. Streeter, "Cost-Benefit Evaluation of Scientific Computing Services", IBM Systems Journal, No. 3, 1972, pp. 219–233.
86. R. Stewart, "Cost Estimating", in Mechanical Engineers' Handbook, Edited by M. Kutz, John Wiley & Sons, Inc., New York, 1986, pp. 1334–1364.
87. J.C. Sureau, "Summary of Results of Antenna Design Cost Studies", 1977, Report No. AD-776 914. Available from National Technical

Information Service, U.S. Department of Commerce, Springfield, Virginia 22151, U.S.A.
88. G.J. Thuesen, "Project Selection and Analysis", in Handbook of Industrial Engineering, Edited by G. Salvendy, John Wiley & Sons, Inc., New York, 1982, pp. 9.4.1.–9.4.27.
89. J.E. Tumarkin, "Statistical Cost Estimating Relationships for Building Construction Planning", Report Study S-318, Institute for Defense Analysis, Arlington, Virginia, 1968.
90. A.M. Tyszkiewicz, "Comparative Cost Model Aircraft Investment Spares", Proceedings of the Annual Reliability and Maintainability Symposium, 1983, pp. 376–382.
91. US News and World Report, "A New Method to Cut Building Costs", US News and World Report, Vol. 71, August 1971, pp. 74–75.
92. T.J. Ward, "Cost-Estimating Methods", American Institute of Chemical Engineers Modular Instruction Series G: Design of Equipment, Volume 1: Plant Design and Cost Estimating, Series Editor: James Beckman, 1986, pp. 12–21.
93. T.J. Ward, "Profitability Analysis", American Institute of Chemical Engineers Modular Instruction Series G: Design of Equipment, Volume 1: Plant Design and Cost Estimating, Series Editor: James Beckman, 1986, pp. 22–32.
94. R. Williams, Jr., "Six-Tenths Factor", Aids in Approximating Costs, in Cost Engineering in the Process Industries, Edited by C.H. Chilton and the Staff of Chemical Engineering, McGraw-Hill Book Company, Inc., New York, 1960, pp. 40–41.
95. A. Zanker, "Estimating Cooling Tower Costs from Operating Data", Chemical Engineering, June 12, 1972, pp. 118–120.
96. O.T. Zimmerman, I. Lavine, Chemical Engineering Costs, Industrial Research Service, Dover, N.H., 1950, 419 pp.

A.8 Maintenance Costs

1. S. Areskoug, "Procurement of Material with Low Maintenance Cost", European Organization for Quality Control (EOQC) Seminar, Torino, Italy, 1971.
2. R.E. Bley, "Optimize Your Maintenance Stores", in Cost Engineering in the Process Industries, Edited by C.H. Chilton and the Staff of Chemical Engineering, McGraw-Hill Book Company, Inc., New York, 1960, pp. 321–323.
3. R.J. Coughlin, "Optimization of Spares in a Maintenance Scenario", Proceedings of the Annual Reliability and Maintainability Symposium, 1984, pp. 371–376.

4. C.F. Crosetto, "Aircraft Operations and Maintenance Cost Model", Report No. D6-53570, The Boeing Aerospace Company, Seattle, USA, June 1968.
5. C.F. Crosetto, "B-1A System Cost—Sixty Test Cases", Report No. D206-10057-1, The Boeing Aerospace Company, Seattle, USA, June 1969.
6. F.P. DaSilva, M.P. DiBenedetto, K.D. Le, J.T. Day, "Quantifying Outage Costs of Generating Units—A Guide to Maintenance Planning", Proceedings of the 13th INTER-RAM Conference For the Electric Power Industry, 1986, pp. 373–379.
7. R. DiStefano, "Development of Ship Equipment Maintenance History", Proceedings of the Annual Reliability and Maintainability Symposium, 1987, pp. 268–273.
8. W.R. Downs, "Maintainability Aspects in Maintenance Management", Proceedings of the Annual Reliability and Maintainability Symposium, 1984, pp. 62–66.
9. D. Gleason, "The Cost of Test System Requirements", Proceedings of the Annual Reliability and Maintainability Symposium, 1981, pp. 108–113.
10. J.C. Iverstine, "Controlling the Cost of Maintenance", Chemical Engineering, March 24, 1980, pp. 103–105.
11. D.P. Johnson, "Cost-Risk Optimization of Non-Destructive Inspection Level", Report No. EPRI: 217-1, TR5, Prepared by Failure Analysis Associates, Palo Alto, California, September, 1975.
12. P.S. Johnson, "SST Operations and Maintenance Simulation Model", Report No. D6A 10266-1, The Boeing Aerospace Company, Seattle, USA, September 1966.
13. J.H. Jordan, "How to Evaluate the Advantages of Contract Maintenance", Chemical Engineering, March 25, 1968, pp. 124–130.
14. R.J. Kaplan, J.Y. Lu, R.M. Paulson, "PRAM: A Preliminary Repair Level Decision Analysis Model", Report No. RM-5752-PR, The Rand Corporation, Santa Monica, California, September 1968.
15. A. Lee, J.D. Hlavacek, "The Relationship Between Price and Repair Service for Consumer Durables", Journal of Marketing, Vol. 40, April 1976, pp. 80–82.
16. J.D. Leonard, "What Does Maintenance Cost?", in Cost Engineering in the Process Industries, Edited by C.H. Chilton and the Staff of Chemical Engineering, McGraw-Hill Book Company, Inc., New York, 1960, pp. 324–334.
17. K.E. Marks, R.W. Hess, "Estimating Aircraft Depot Maintenance Costs", The Rand Corporation Report No. R-2731-PA & E, July 1981. Available from the Rand Corporation, Santa Monica, California 90406, U.S.A.
18. D. Marlin, "Keeping an Eye on Maintenance Costs", Process Engineering, Vol. 7, 1978, pp. 47–48.

19. T. Nakagawa, "Optimum Preventive Maintenance Policies for Repairable Systems", IEEE Transactions on Reliability, Vol. R-26, 1977, pp. 168–173.
20. F.S. Nowlan, "Reliability-Centered Maintenance", in Design to Cost and Life Cycle Cost, AGARD Conference Proceedings No. 289, May 1980, pp. 16-1 to 16-13. Available from National Technical Information Service, Springfield, Virginia, U.S.A.
21. R.H. Oggett, "Terotechnology—Its Practical Applications in Airline Maintenance", Terotechnica, Vol. 1, No. 4, July 1980, pp. 265–269.
22. M. Phister, Jr., "Analyzing Computer Technology Costs—Part 2: Maintenance", Computer Design, October 1978, pp. 109–118.
23. H.L. Platzer, "A Maintenance Philosophy for Mainframe Computers", Proceedings of the Annual Reliability and Maintainability Symposium, 1984, pp. 359–361.
24. M.F. Reid et al., "Design—The Requirements of Maintenance", IEE Proceedings, Part A, Vol. 131, No. 8, November 1984, pp. 646–653.
25. D.J. Smith, A.H. Babb, Maintainability Engineering, John Wiley & Sons, New York, 1973.
26. R.F. Steiner, "A Simple Method to Reduce Avionics Repair Costs", Proceedings of the Annual Reliability and Maintainability Symposium, 1983, pp. 432–434.

A.9 Operating Costs

1. N.E. Betaque, Jr., M.R. Fiorello, "Aircraft System Operating and Support Costs: Guidelines for Analysis", Report No. AD A039369, April 1977, 94 pp. Logistics Management Institute, 4701 Sangamore Road, Washington, D.C. 20016, U.S.A.
2. R.E. Cavender, "Application of the A-X 10-year Operating and Support Cost Model to the F-4E", Operations Analysis Technical Memorandum No. 14, Air Force Logistics Command, Wright-Patterson Air Force Base, Ohio, USA, March 1971.
3. R.E. Cavender, "Variability of Measured Logistic Effects as Computed via the A-X 10-year Operating and Support Cost Model", Supplement No. 1 to Operations Analysis Technical Memo No. 14, Air Force Logistics Command, Wright-Patterson Air Force Base, Ohio, USA, October 1971.
4. L.J. Dierking, G.L. Foster, P.N. Dyer, "Fuel Quality Characteristics and Their Impact on Operating Costs", Proceedings of the 13th INTER-RAM Conference for The Electric Power Industry, June 2–5, 1986, pp. 358–364.
5. D.G. Downing, "Calculating Minimum-Cost Ion-Exchange Units", in Modern Cost-Engineering Techniques, Edited by H. Popper and the Staff

of Chemical Engineering, McGraw-Hill Book Company, New York, 1970, pp. 472–476.
6. M.R. Fiorello, L.G. Jones, Jr., "Combat Vehicle System Operating and Support Costs: Guidelines for Analysis", Report No. AD A041508, June 1977, 100 pp. Logistics Management Institute, 4701 Sangamore Road, Washington, D.C. 20016, U.S.A.
7. K. Grayson, "A New Method for Estimating Transport Aircraft Direct Operating Costs", in Design to Cost and Life Cycle Cost, AGARD Conference Proceedings No. 289, May 1980, pp. 14-1 to 14-20. Available from the National Technical Information Service (NTIS), Springfield, Virginia, U.S.A.
8. R.E. Houts, O & S Cost Visibility in Early Design", in Design to Cost and Life Cycle Cost, AGARD Conference Proceedings No. 289, May 1980, pp. 2-1 to 2-12. Available from National Technical Information Service (NTIS), Springfield, Virginia, U.S.A.
9. J.H. Keeney, "General Operations and Logistics Simulation (Goals): Description and Results", Report No. D162-10155-1, The Boeing Aerospace Company, Seattle, USA, December 1969.
10. S.L. Lazo, "The cost of RAM: An Analysis of Operating and Maintenance Expenditures", Proceedings of the 13th Inter RAM-Conference For The Electric Power Industry, June 2–5, 1986, pp. 350–357.
11. F.P. O'Connel, "Chart Gives Operating Labor for Various Plant Capacities", in Modern Cost Engineering Techniques, Edited by H. Popper and the Staff of Chemical Engineering, McGraw-Hill Book Company, New York, 1970, pp. 251–252.
12. P.R. Walton, "Sources of Error in Operating-Cost Estimates", in Modern Cost-Engineering Techniques, Edited by H. Popper and the Staff of Chemical Engineering, McGraw-Hill Book Company, New York, 1970, pp. 240–242.
13. E.L. Wienecke, E.E. Feltus, "Predictive Operations and Maintenance Cost Model, Volume I", Report No. AD A078052, August 1979, 282 pp. Air Force Avionics Laboratory, System Evaluation Group (AFAL/AAA-3), Wright-Patterson AFB, Ohio 45433, USA.

A.10 Miscellaneous

1. J. Bateson, "In-Circuit Testing", Van Nostrand Reinhold Company, New York, 1985, pp. 55–61, 197–218.
2. E.F. Bornholtz, "Company and Cost Accounting", in Handbook of Industrial Engineering, Edited by G. Salvendy, John Wiley & Sons, New York, 1982, pp. 9.1.1.–9.1.20.
3. J.R. Canada, Intermediate Economic Analysis for Management and Engineers, Prentice-Hall, Englewood Cliffs, New Jersey, 1971.

4. G.D.C. Chiang, "Mock Trial; Goody et al. vs. Big George Motor Company", Proceedings of the Annual Reliability and Maintainability Symposium, 1979, pp. 83–84.
5. G.E. Dieter, "Engineering Design: A Materials and Processing Approach, McGraw-Hill Book Company, New York, 1983.
6. W.J. Dodge, "Optimum Insulation Thickness—by Computer", in Modern Cost Engineering Techniques, Edited by H. Popper and the Staff of Chemical Engineering, McGraw-Hill Book Company, Inc., New York, 1970, pp. 477–480.
7. L.R. Dorsky, "Management Measures for Customer Satisfaction", Proceedings of the Annual Reliability and Maintainability Symposium, 1985, pp. 13–16.
8. S.R. Dwyer, "Reliability Targets for a Calibration Program", Proceedings of the Annual Reliability and Maintainability Symposium, 1984, pp. 367–370.
9. A.J. Eccleston, "Cost Considerations of Engine Fuel Control Systems", in Design to Cost and Life Cycle Cost, AGARD Conference Proceedings No. 289, May 1980, pp. 27-1 to 27-14. Available from National Technical Information Service (NTIS), Springfield, Virginia, U.S.A.
10. H.A. Eimstad, Jr., "Reliability and Maintainability Technical and Cost Relationships", Proceedings of the Annual Symposium on Reliability, 1968, pp. 331–337.
11. H. Elgee, "Investing in Major Spare Parts", in Modern Cost Engineering Techniques, Edited by H. Popper and the Staff of Chemical Engineering, McGraw-Hill Book Company, 1970, pp. 493–496.
12. H.E. Figgie, Jr., "Product Redesign as a Cost-Cutting Tool", Machine Design, June 26, 1986, pp. 81–84.
13. J. Garfield, I. Bazovsky, Jr., "Economical Fault Isolation Analysis", Proceedings of the Annual Reliability and Maintainability Symposium, 1985, pp. 480–484.
14. M. Girand, "Problems in Optimal Spare Provisioning", NATO ASI Series (Series F: Computer and Systems Sciences, No. 3), Edited by J.K. Skwirzynski, Springer-Verlag, Munich, 1983, pp. 655–682.
15. H. Hecht, "Figure of Merit for Fault-Tolerant Space Computers", IEEE Transactions on Computers, Vol. C-22, No. 3, March 1973, pp. 246–251.
16. D.C. Hittle, "Use of the Building Loads Analysis and System Thermodynamics Program to Perform Total Energy System Analysis", Construction Engineering Research Lab., U.S., June 1977, 19 pp., NTIS No. AD-A040 744/5ST. Available from the National Technical Information Service, Springfield, Virginia.
17. L.A. Hoffman, "The Team Approach to Products Liability Prevention and Defense", Proceedings of the Annual Reliability and Maintainability Symposium, 1985, pp. 6–12.

18. L.A. Hopkins, "Availability Cost/Benefit System Design Criteria", Proceedings of the Reliability Conference for the Electric Power Industry, 1978, pp. 111–118.
19. S.A. Hovanessian, "Research and Development of a Large-Scale Electronic System", IEEE Transactions on Engineering Management, Vol. EM-22, No. 3, August 1975, pp. 94–101.
20. P.A. Hudak, K.A.B. Macdonald, "New Technology Transport for Fuel Critical Economy", Proceedings of the Annual Reliability and Maintainability Symposium, 1980, pp. 101–105.
21. R.M. Jacobs, "Design Review: A Liability Preventer", Mechanical Engineering, Vol. 97, No. 8, August 1975, pp. 34–39.
22. I.W. Kabak, "System Availability and Some Design Implications", Operations Research, Vol. 6, 1980, pp. 349–360.
23. C.H. Karr, G.L. Wagner, "R&M—Today's Heating and Cooling vs. Solar Energy", Proceedings of the Annual Reliability and Maintainability Symposium, 1976, pp. 491.
24. L. Klein, "Models Differ Mostly in Terms of Doctrine", Mechanical Engineering, Vol. 109, No. 1, January 1987, pp. 53–55.
25. J.P. Kornfeld, "Cost-Effectiveness as a Method of Evaluating Design Changes", IRE Transactions on Engineering Management, Volume EM-9, No. 1, March 1962, pp. 15–23.
26. M. Lenard, "Comparing Guideway Transit Implementation Plans, Including the Impact on Road Traffic", Transportation Planning Technology, Vol. 6, No. 1, 1980, pp. 33–43.
27. W. Leontief, "The Ins and Outs of Input/Output Analysis", Mechanical Engineering, Vol. 109, No. 1, January 1987, pp. 28–35.
28. B. Livson, I. Melamed, S. Samuel, "Optimization of Reliability for Mini-RPV", Proceedings of the Annual Reliability and Maintainability Symposium, 1982, pp. 30–33.
29. R.T. Lund, "Making Products Live Longer", Technology Review, Vol. 79, No. 3, January 1977.
30. E.M. Malstrom, "An Overview of Cost Engineering", in Manufacturing Cost Engineering Handbook, Edited by E.M. Malstrom, Marcel Dekker, Inc., New York, 1987, pp. 1–66.
31. D.M. Marko, and T.D. Schoonmaker, "Optimizing Spare Module Burn-In", Proceedings of the Annual Reliability and Maintainability Symposium, 1982, pp. 83–86.
32. L.S. Mickel and E.E. Newman, "Sparing to Optimize Naval Airship Availability", Proceedings of the Annual Reliability and Maintainability Symposium, 1987, pp. 264–267.
33. P.M. Paige, "Costlier Cooling Towers Require a New Approach to Water-Systems Design", in Modern Cost-Engineering Techniques, Edited by H. Popper and the Staff of Chemical Engineering, McGraw-Hill Book Company, New York, 1970, pp. 451–455.

34. H.A. Quigley, "Economics of Multiple Units", in Modern Cost-Engineering Techniques, Edited by H. Popper and the Staff of Chemical Engineering, McGraw-Hill Book Company, Inc., New York, 1970, pp. 421–426.
35. F.R. Riddell, R.A. Gunkel, "An Integrated Computer-Aided Logistics Support System", Proceedings of the Annual Reliability and Maintainability Symposium, 1986, pp. 477–479.
36. J. Rose, "Fault-Tolerant System Optimization", Proceedings of the Annual Reliability and Maintainability Symposium, 1980, pp. 95–100.
37. H.E. Schweyer, "Your Process Economics", in Cost Engineering in the Process Industries, Edited by C.H. Chilton and the Staff of Chemical Engineering, McGraw-Hill Book Company, Inc., New York, 1960, pp. 449–452.
38. P.E. Seibert, "Structural Integration as a Means of Cost Reduction", in Design to Cost and Life Cycle Cost, AGARD Conference Proceedings No. 289, May 1980, pp. 10-1 to 10-17. Available from the National Technical Information Service, Springfield, Virginia, U.S.A.
39. A.A. Seireg, "Establishing the Market Value of Innovation", Mechanical Engineering, January 1987, pp. 36–39.
40. J.E. Snell, P.R. Achenbach, S.R. Petersen, "Energy Conservation in New Housing Design", Science, Vol. 192, No. 4246, June 25, 1976, pp. 1305–1311.
41. E.E. Stiepel, "Product Liability-Considerations in Machinery Design", Mechanical Engineering, Vol. 104, No. 5, May 1982, pp. 64–68.
42. D.N. Streeter, "Productivity of Computer-Dependent Workers", IBM Systems Journal, Vol. 14, No. 3, 1975, pp. 292–305.
43. C. Suntag, "Product Liability Prevention Program", American Society for Quality Control (ASQC) Quality Congress Transactions, Boston, 1983, pp. 396–403.
44. H.G. Thuesen, W.J. Fabrycky and G.J. Thuesen, Engineering Economy, Prentice-Hall, Englewood Cliffs, New Jersey, 1977.
45. E.S. Winlund, "Cost-Effectiveness Analysis for Optimal Reliability and Maintainability", Proceedings of the 11th National Symposium on Reliability and Quality Control, 1965, pp. 107–114.

Index